# PLANETS AND LIFE

THE WORLD OF SCIENCE LIBRARY
GENERAL EDITOR: ROBIN CLARKE
EDITOR OF SCIENCE JOURNAL

# PLANETS AND LIFE

P. H. A. Sneath

FUNK AND WAGNALLS     NEW YORK

Library of Congress Catalog Card Number: 70-93940

First American edition published in 1970 by arrangement
with Thames & Hudson International Ltd. (London)

Funk & Wagnalls, *A Division of* Reader's Digest Books, Inc.

Printed in the Netherlands

# CONTENTS

PREFACE                                              7

1  THE PHYSICAL UNIVERSE                             9
   The changing perspective                          9
   The origin of galaxies                           18
   From quasars to quarks                           22

2  TERRESTRIAL LIFE                                 25
   The generation of life                           26
   The chemical unity of life                       31
   The genetic code                                 34
   The variety of living things                     38
   Conditions for survival                          41
   Longevity and dormancy                           48

3  THE NATURE OF LIVING SYSTEMS                     55
   The concept of living                            55
   Organism and organization                        61
   The information content of organisms             64
   The physical medium of life                      70

4  THE ORIGIN OF TERRESTRIAL LIFE                   73
   The panspermia hypothesis                        74
   Abiogenesis                                      75
   The dawn of life                                 78
   The evolution of higher organisms                82

5  LIFE BEYOND THE EARTH                            89
   The minor planets                                90
   The earth                                        96
   The planet Mars                                  97
   The outer planets                               105
   The moon                                        108
   Satellites and asteroids                        113
   Organic matter in meteorites                    114
   Life-detection apparatus                        118
   The probability of alien life                   124

## 6 ALTERNATIVE BIOCHEMISTRIES 131
One or many recipes? 131
Mirror creatures 133
Vital solvents 134
Ammonia life 135
Why carbon? 138
Silicon salamanders 141

## 7 INTELLIGENT LIFE 144
Bug-eyed monsters 144
Unfamiliar modes of perception 148
The evolution of intelligence 150
Could machines live? 156
Consequences of Good's Law 161
Human biomachines 164

## 8 SPACE TRAVEL 167
Exploring the solar system 167
Journeying to the stars 176
The aging of astronauts 177
Problems of contamination 180
The threat to life on Earth 182

## 9 INTERSTELLAR COMMUNICATION 189
The Order of the Dolphin 189
Chances of making contact 191
Project Ozma 196
Interstellar languages 200
Flying saucers 201

## GLOSSARY 205

## BIBLIOGRAPHY 209

## SOURCES OF ILLUSTRATIONS 211

## INDEX 214

# PREFACE

We live on the threshold of an era of space travel, but what this new era will bring is only dimly foreseen. The radio astronomer Otto Struve, in stating his belief that intelligent life exists elsewhere in the universe, has said that in addition to the classical laws of physics we may now have to take into account the role of intelligent beings—aliens as well as ourselves—in shaping events in the universe at large. This idea, he suggests, may yet rank with the Copernican revolution in transforming our view of the cosmos.

Prefaces are seldom read, and I am content to make only one further point: the reader may feel that some topics are given undue emphasis. But as J. B. S. Haldane has pointed out, this is necessary if people are to be encouraged to think about them. I do not apologize for referring to science fiction. SF writers have much the same function as seers of old: to make us ask ourselves where we are going. When we stop asking questions it is not because we know the answers, but because we stop trying.

Acknowledgements to sources of information are given elsewhere, but I should like to mention here that although the views expressed in this book are my own, much inspiration derives from the work of Haldane, Joshua Lederberg, Philip Morrison and Carl Sagan, to each of whom I am indebted.

*P.H.A.S.*

*Medical Research Council*
*Microbial Systematics Research Unit,*
*University of Leicester, 1969*

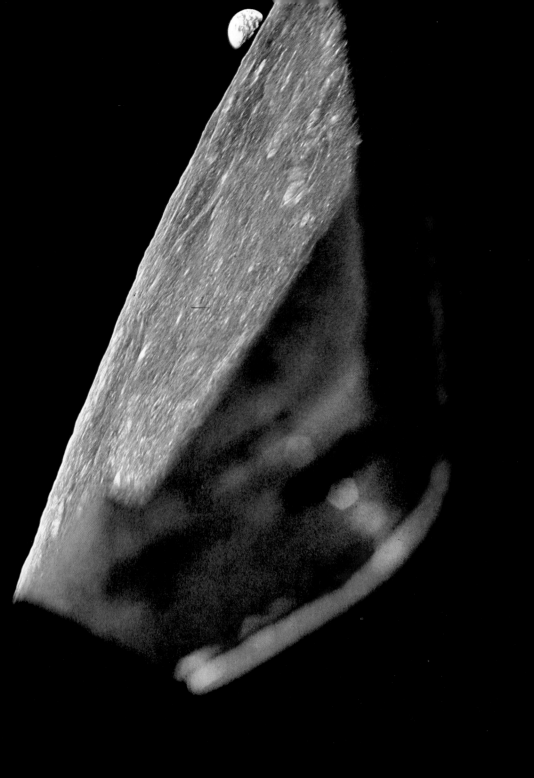

# THE PHYSICAL UNIVERSE 1

In December 1968, three American astronauts orbiting the moon became the first men to view the earth from the vicinity of another celestial body. This signified more than the final demise of the Flat Earth Theory. It was the first time that man in person had viewed the universe from anywhere but the earth. Although our vantage point has until so recently been restricted to the globe on which we live, over the centuries our ideas of the universe have undergone profound changes.

## The changing perspective

When man first began to observe the stars, probably in the late Palaeolithic stage of his development, about half a million years ago, it seemed natural to think of the universe as a flat piece of land bounded by the sea and domed by a transparent heaven in which were set the sun, the moon and the stars. The sun and moon have a motion apparent to even the most casual eye, but it would have been rather later that men noticed certain 'stars'—the planets of our solar system—also moving in the heavens against the background of fixed stars. All celestial objects—which would also include occasional comets (omens of disaster, the 'hairy stars') and the more frequently observed meteor trails ('shooting stars')—were thought to be relatively close to the earth. In so far as they were viewed physically at all—rather than as magical objects personified as

*Opposite: a life-bearing planet on a lunar horizon. This evocative view of the earth from a space craft in moon orbit was photographed during the historic Apollo-8 mission of December 1968*

gods and spirits—they were thought of as lying on the inside of the crystal dome of the heavens, their distance being measured in terms of terrestrial distances—in hundreds of miles. Beneath the earth was the ocean, on which all floated.

The astronomers of Mesopotamia and India are the earliest of whom we have evidence. They were the first to record celestial phenomena systematically, although the intimate links between the motions of the sun and moon and the proper times for seed sowing and harvest must have been of critical interest to the first farmers at the beginning of the Neolithic period (*c.* 10,000 BC). But it was the Greeks who first attempted to explain astronomical phenomena in physical terms. One of the earliest of the Greek wise men, Thales of Miletus (*fl.* 600 BC), was celebrated for predicting the solar eclipse that occurred during the battle between the Lydians and the Medes in 585 BC. The followers of Pythagoras in the sixth century BC were apparently the first to conceive of the earth as a globe freely poised in space. Greek astronomy culminated, in the second century AD, in the system of Ptolemy, in which the heavenly bodies were held to rotate about the earth. Already the scale of the universe had grown considerably. Around 200 BC Eratosthenes had calculated the size of the earth with remarkable accuracy (it was now a globe and no longer flat), and had realized that the starry heavens were many millions of miles away. As some guide to the scale of the universe which the ancients contemplated, we may note that the great Archimedes, in his work *The Sand Reckoner*, suggests that the universe might contain $10^{63}$ sand grains, which would give a diameter of the order of thousands of millions of miles.

The Copernican revolution in the sixteenth century AD was of great importance in the development of science. The sun, not the earth, became the centre of the universe. Never again could men think of themselves as unique, as the only possible sentient beings. At least in theory, other earths could exist. Indeed, the sixteenth-century philosopher Giordano Bruno said

*The recently observed comet Ikeya-Seki hurtles through the night sky at some 300 miles per second as it dips towards the sun. Such phenomena were once regarded with foreboding as signs from the gods*

they did, and that they were peopled by intelligent creatures; and Bruno was one of the first (perhaps anticipated by Thomas Digges) to portray the stars, not as being fixed to a crystal sphere, but as scattered throughout infinitely deep space. The new star seen by Tycho Brahe—the stellar explosion, or 'supernova', of 1572 showed that the crystal firmament of stars was not unchanging, while the crystal spheres themselves were soon to vanish under the scrutiny of Galileo's telescope. Galileo showed that the Milky Way consists of innumerable faint stars, extending to previously unimagined distances. Man's conception of the scale of the universe was rapidly expanding.

In the late seventeenth century, Isaac Newton, by means of his theory of gravitation, linked the movements of the planets together into one physical system, making the point that the *same* natural laws control the earth and the stars. Until his time quite separate laws were thought to rule the heavens and the earth. At about this time, too, the sun was shown to be

*Scholars of the Middle Ages, closer to Roman than Greek sources, upheld the idea of a flat earth surmounted by a celestial vault. The motions of sun, moon, planets and stars were attributed to mysterious external agencies*

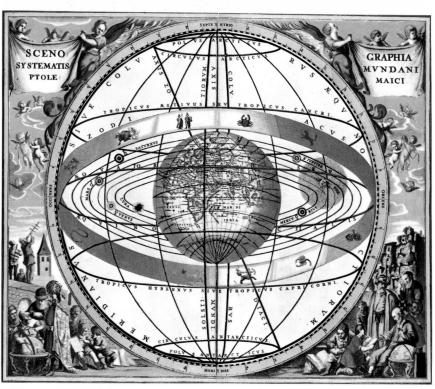

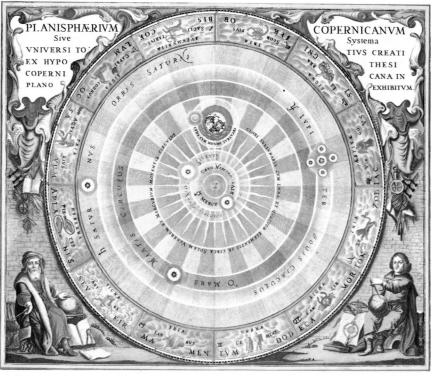

almost 100-million miles away from the earth, and the solar system accordingly to be thousands of millions of miles in diameter.

The distances to the stars nevertheless remained uncertain. Although the sun itself was now realized to be a common star, even the brightest stars in the night sky showed no measurable disc in the best telescopes. The parallax method of measuring the minute changes in position of the nearer stars against the background of the more distant ones when viewed from the earth at opposite sides of its yearly orbit was long in coming. At last, in 1838, the German astronomer F. W. Bessel was able to give the astonishing news that even the nearest stars are some tens of millions of millions of miles away. If the sun were the size of an apple, the nearest stars would still be thousands of miles away.

Distances of millions of millions of miles mean very little, and they are inconvenient units to work with. Distances to the stars are therefore measured in *light-years*, the distance that light travels in one year. The speed of light is 186,000 miles per second, so a light-year is 5,880,000,000,000 miles. Another measure used for

*Opposite top: a seventeenth-century Dutch cartographer's version of the Ptolemaic universe, with the earth poised at its centre. Bottom: a matching account of the Copernican system—the first to recognize the earth as a satellite of the sun. Below: William Blake's symbolic portrayal of Isaac Newton whose formulation of the law of gravity enabled the motions of the planets to be described in relatively simple mathematical terms*

*Below: an unusual view of the great 200-inch Hale reflector at Mount Palomar, California. Astronomers have a choice of three observation positions, one of which (as can be seen here) is actually inside the instrument. Opposite: an example of what such a telescope may reveal: a spiral galaxy in the constellation Ursa Major. This 'island universe', comprising many millions of stars, is typical of the millions upon millions of similar formations now known to lie beyond our own galaxy, out to the farthest limits of observation*

these distances is the *parsec*, which is 3·26 light-years.

The discovery of these vast distances between the stars came as a shock to many. The earth, far from being of central importance, was no more than a minor planet of an average star. But this was only the beginning. Astronomers had long been puzzled by the faint wisps of light usually showing spiral markings and known as the spiral nebulae. These showed no evidence of stars in the early telescopes, but were clearly at a great distance. The suggestion of G. W. Ritchey and H. D. Curtis, based on the observation of stellar explosions—supernovae—in several of the spiral nebulae was that these nebulae were millions of light-years away. Few astronomers were willing to accept these enormous figures. At last the great reflecting telescopes of the Mount Wilson and Palomar Observatories and the pioneer work of Edwin P. Hubble,

showed that stars were present in the nearest of the spiral nebulae, among them the Great Nebula in Andromeda. Ingenious methods were found of estimating the true brightness of some of these stars, and thus (by comparing this with the apparent brightness) of estimating their distance from us. This showed yet another astounding extension of the scale of the universe. The nearest spiral nebulae (now called galaxies) were 2-million light-years away. It became clear that the Milky Way itself was simply another galaxy, a flattened disc of stars and gas, more than 100,000 light-years across and about 2,000 thick. Our solar system is near its edge, some 30,000 light-years from the galactic centre. Wherever astronomers could look out past the dust and clouds of our own galaxy, represented mainly by the Milky Way, they saw thousands upon thousands of galaxies. The most distant yet observed are over 1,000-million light-years away.

These distances are as great as any we are ever likely to know about, since the galaxies appear to be receding from us at great speeds. This is shown by the reddening of the light that they emit. This 'red shift' shows, too,

*Lines in the spectra of light emitted by receding galaxies are displaced towards the lower-frequency, or red, end of a standard reference spectrum, the degree of 'red shift' depending on the velocity of recession. The indicated velocity in the example below is 86,000 miles per second—nearly half the speed of light*

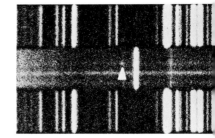

that the most distant are receding from us at the greatest speeds. At distances much greater than these the recession would be close to the speed of light, and light from them would reach us in so reddened a form that we would scarcely recognize it.

The development of radio-astronomy, from the pioneering days of K. G. Jansky to the coming of the vast bowls of Jodrell Bank and Arecibo, has greatly extended our knowledge, for it has allowed us another glimpse through the barrier of our atmosphere, which only allows a tiny fraction of radiation to reach us. Radio-astronomy has revealed some extraordinary new objects whose nature is still very poorly understood. Radio waves of 21 cm. wavelength are emitted by hydrogen gas, and this can be used to map the gas clouds in our galaxy. Another source is the turbulent remains of stellar explosions, for example the Crab

*The radio telescope confronts astronomers with a universe of radio sources complementing the universe of light sources accessible to the optical telescope. Opposite: the Parkes radio telescope in New South Wales, Australia. Below: a light source associated with strong radio emissions: the Crab Nebula, the remains of a star which was seen to explode in AD 1054*

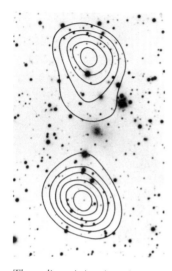

*The radio-emission intensity contours of one of the strongest radio sources in the sky. This source, known as 'Cygnus A', is associated with a faint but visible galaxy lying midway between the radio areas*

*Quasars—quasi-stellar radio sources—such as the one shown below, shining as brightly as 200 average galaxies, remain the most baffling of celestial objects yet observed*

Nebula. At least one intense radio source, Cygnus A, is apparently due to a collision between two galaxies some 500-million light-years away.

But the most puzzling objects are those known as quasars (an abbreviation for quasi-stellar radio sources). These are intense sources of both radio and light waves. They are much too small to be galaxies, but appear to be at very great distances (though this is still disputed). They emit enormous amounts of energy, and are brighter than all the stars in a galaxy put together. Another group of radio sources are stars known as pulsars which pulsate quite regularly, and will be mentioned again later. They are small objects and not very far away by cosmic standards.

## The origin of galaxies

We now have a view of the universe very different from that of our ancestors. The diameter of the known part of it is measured in thousands of millions of light-years. The galaxies which it comprises are thousands of light-years in diameter, and are flattened masses of gas and stars, slowly rotating around the galactic centre, and taking many millions of years to complete one revolution. The galaxies show a spiral structure, and one of them may contain as many as a 100,000-million individual stars. Most of the universe is empty space: even within the galaxies the concentration of matter is very low, so that light traverses the universe with little hindrance. We have seen that the galaxies appear to be receding from us, and the most distant are receding at a speed of one-third the velocity of light.

This raises some interesting problems about the origin of the universe. We may first note that we see the more distant galaxies as they were many millions of years ago, because it has taken this time for their light to reach us. Also, if we extrapolate backwards from their present positions, we find that about 10,000-million years ago the galaxies would all have been contained in a rather small volume of space. The universe would then have consisted of matter and radiation at a very high temperature, perhaps of the order of 15,000-million°C, and it would rapidly expand, this expansion continuing to the present day. This has suggested the 'Big Bang' theory of the origin of the universe. The Belgian scientist Georges Lemaitre supposed that the universe was created as a single primeval atom which was unstable and at once disintegrated. It has also been proposed that the universe is pulsating, alternating between an expanding state and a contracting state, with the latest period of extreme contraction having occurred about 10,000-million years ago. The points of distinction between the Big Bang and 'Oscillating' universes are quite esoteric, and for our purposes we can group them together.

The other main theory is the 'Steady State' theory of the British astronomers Hoyle, Bondi and Gold. In this theory the steady creation of matter is postulated. The new matter 'feeds' the galaxies with what is presumed to be hydrogen gas, and also gradually forms new galaxies in space. In terms of such a model the galactic recessions could be maintained indefinitely, and it would not be necessary to propose a finite starting time.

It is not easy to distinguish experimentally between the Big Bang and Steady State models. However, several tests have been proposed of which two are readily explained. The first turns on the following argument: If the Steady State theory is correct then all parts of space will be much the same and will contain a mixture of young and old galaxies. But if the Big Bang theory is correct, the most distant galaxies will look younger than the average, because we see them as they were long ago, owing to the greater time it has taken for their light to reach us. A similar test can be made with radio waves detected by the newest weapon in the armoury of the cosmologist, the radio telescope. As radio waves from the most distant galaxies are recorded, new evidence is slowly accumulating, and the weight of this evidence is against the Steady State theory. The second test (which also requires the radio

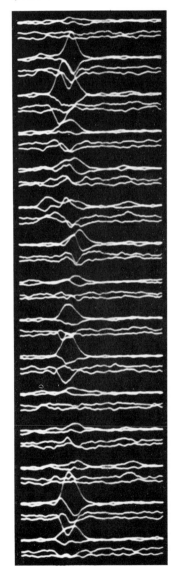

*Traces of a regular radio pulse emitted by a pulsar—another kind of radio star*

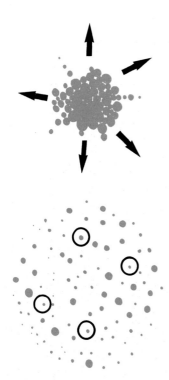

*Two different views of the expanding universe. The Big Bang theory (top) holds that the density of the universe as a whole is steadily decreasing, since the total amount of matter remains constant. But according to the Steady State theory (above) new matter, 'created' in the gaps left by the expansion, keeps the overall density constant*

*Opposite: a potential birthplace of further stars: the great dust and gas cloud of the Orion Nebula, suffused with the light of four extremely hot stars embedded within it*

telescope) involves the residual radiation from the original Big Bang. As the universe expands, this radiation, which in the primeval fireball was gamma radiation, would become longer and longer in wavelength, until now it would be almost entirely in the form of radio waves. These waves would have two important properties; their intensity would be the same in all directions, and they would show a spectrum of the kind called 'black-body radiation'. This radiation has recently been detected, and shows both of these critical properties. Since black-body radiation of this kind could not arise in a Steady State universe this finding again supports the alternative hypothesis. The origin of the different chemical elements is perhaps most readily explained by some form of Big Bang. Most astronomers now favour either the Big Bang or some form of oscillating universe, in which perhaps the contracted state consists mainly of radiation and the expanded state mainly of matter. The age of the universe (or at least the time from the latest contraction) is still in dispute, with estimates ranging from 7,000- to 12,000-million years.

The age of the matter that goes to make up the earth, and also the meteorites, is, however, fairly well established at around 4,600-million years, an estimate based on radioactivity measurements. The argument runs on the following general lines: If the earth were much older than 4,600-million years, then very little of the radioactive elements like uranium, rubidium-87 and potassium-40 would now remain, while their breakdown products (certain isotopes of lead, strontium, calcium and argon) would be more abundant than they are. If the earth were much younger than this, there would be more of the radioactive elements than we observe. Since the different radioactive isotopes have different rates of decay, there is only a narrow range of times that fits all the evidence, and this is between four- and five-thousand million years.

The new knowledge of the galaxies shows that stars are forming out of the tenuous gas that lies between them in clouds. It is primarily gravitation which pulls

in this gas (mostly hydrogen) and it is gravitation again which compresses the gas until it becomes so hot that thermonuclear reactions begin at the star's centre. These provide the great bulk of the star's energy, including that of our sun. The thermonuclear reactions also produce heavier elements. It is not entirely clear how the matter of the earth was formed. Some believe that the presence of the heaviest elements indicates that it was produced in the intense reaction of a supernova explosion, about 4,500-million years ago. Others favour the theory that it was formed in the Big Bang. One day, perhaps, we shall be able, by analysing the light emitted, to determine the different isotopes in distant stars and galaxies with enough accuracy to answer these questions.

But wherever the matter itself came from, how did our own solar system form? The theory, held for many years, that the planets condensed from a fiery tongue of the sun's matter, pulled out by gravitation on the near passage of another star, is now quite out of favour. For one thing, the chances of this happening are minute, and yet there is evidence that planetary systems are not uncommon. For another, the sun rotates much too slowly to fit in with this theory. It is now generally believed that the solar system was formed by the accretion of particles within a cloud of cold dust and gas, mostly hydrogen, but containing heavier elements as well. The bulk formed the sun, the rest forming the planets. This agrees with the large amount of hydrogen in the giant planets like Jupiter, which were heavy enough to attract and retain most of the gas that was left over from the sun. The smaller planets had less gravitational pull, and got and retained less of the light elements like hydrogen.

### From quasars to quarks

While astronomers and astrophysicists were exploring the frontiers of the universe, other physicists were turning their attention to the very small, to the structure of the atom, and this, too, bears on our theme of cosmic biology. It was no accident that some, like Albert

Einstein, were active in both fields, because the properties of light, gravity, electric charge and so on are of interest in both very large and very small phenomena. The crowning achievement of this work was perhaps the realization of the equivalence of mass and energy, summed up in Einstein's famous equation: $e = mc^2$, where $e$ is energy, $m$ is mass and $c$ is the velocity of light.

But in exploring the structure of the atom some very surprising things were discovered. It was not simply that the atom is after all divisible, as the British physicists J. J. Thompson and Lord Rutherford showed; nor that the atom consists of a small positively charged heavy nucleus surrounded by negatively charged electrons, very much like planets round the sun. What was surprising was that the ordinary laws of motion do not apply on the atomic scale of size, contradicting Newton's assumption that the same laws apply to all situations. (We now know, too, that Newtonian physics does not hold exactly on the largest scale either, as Einstein's Theory of Relativity shows, but this need not concern us here.)

At the atomic level something very odd was happening. If an electron simply orbited the nucleus like a planet round the sun, it would lose energy and, within a fraction of a second, should stop as it spirals in toward the nucleus. Yet electrons evidently do not behave like this. Instead they have certain permitted orbits, and can only occupy these. The reason is that atomic particles can only gain or lose energy in fixed packets or 'quanta', and on this has been built the elegant theory of quantum mechanics. If an electron jumps from one orbit to another it emits or absorbs a quantum of energy in the form of a packet of electromagnetic radiation, a photon (which can be anything from a low-energy wireless wave to a high-energy X-ray or gamma ray, with light waves in between). The electron is never intermediate in position, for it 'leaps' from one orbit directly into another.

These electron orbits are important to our theme because they confer on the atom the properties of a

*A field-ion micrograph of the tip of a tungsten wire. The bright 'scintillation spots' represent the sites of individual atoms*

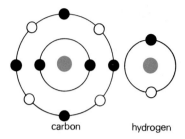

carbon            hydrogen

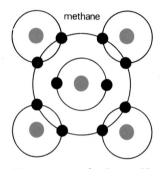

methane

*How one atom of carbon is able to combine with four atoms of hydrogen. The blue discs represent atomic nuclei, the black discs electrons, and the white available spaces for additional electrons*

particular chemical element. They determine what sort of chemical compounds it can form and what chemical reactions it can undergo. The number of electrons is ordinarily fixed, because the number of positive charges in the atomic nucleus is the same as the number of electrons round it. The possible orbits, too, are greatly restricted to certain 'shells', as are the numbers of electrons that can occupy any one of them. One element in particular, carbon (C), has an unusual ability to form complex compounds with other atoms. The carbon atom has 2 electrons in its inner shell, and in its second shell has 4 electrons (out of a possible maximum of 8, which is the case with the neon atom). These 4 outer electrons are readily shared with other atoms, and this sharing of electrons serves to bind the other atoms firmly to the carbon atom. Carbon is particularly adept at sharing its electrons, either with other elements or with other carbon atoms. It can gain 4 electrons from other atoms to achieve the stable condition of 8 electrons in its outer shell, and this, incidentally, confers on it its 'valency' of 4 which acts like a set of arms to link up carbon atoms into complex patterns of molecules. These properties confer on carbon the ability to form large and very complex molecules. Life as we know it is based on these carbon compounds.

The diameter of an atom is around a hundredth of a millionth of a centimetre, while the diameter of the hydrogen nucleus is about a tenth of a million-millionth of a centimetre. The elementary particles that make up the atomic nucleus cannot very readily be measured in size units, because at these minute distances they do not behave in quite the way we expect. Modern physics has demonstrated the existence of a large number of 'smaller' subatomic particles, some of them very short-lived, and it has recently been suggested that they are all composed of yet 'smaller' particles, called quarks. These are the 'smallest' things man has tried to study, and may be contrasted with the observed diameter of the universe, about $10^{41}$ times as large.

# TERRESTRIAL LIFE

In contrast to the physical universe, the world of life has little to do with the very big or the very small. It is, rather, the world of the *very complex*. Our knowledge of biology has progressed from a fairly simple view of living things to an awareness of their delicate and involved construction, and their subtle interaction with other living things. Indeed, in the latter respect we may not have progressed much. Our views of the balance of nature—of foods and pests, of pest-killing and of exploiting the land—have not changed very much since the dawn of man. We have learned a great deal, however, about the complexities of living tissues and their hereditary endowment. In large part this new knowledge was revealed by the study of the very small. The microscope played the same part in extending biological knowledge that the telescope played in astronomy. Anton van Leeuwenhoek's glimpse of the swarming life in a drop of pondwater was the analogue of Galileo's view of the Milky Way. The microscope revealed a new world to the fascinated eyes of natural historians. It showed, too, that all but the simplest organisms were composed of great numbers of cells, each with a nucleus within it.

The growth of biology was much more erratic than that of astronomy. With the exception of Aristotle's descriptive work on animals and that of the early herbalists on plants, little progress was made until late in the Renaissance. Such studies of anatomy and

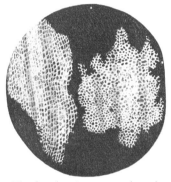

*The first drawing ever to show the cellular structure of living matter. It was made by Robert Hooke in 1665 while examining cork samples under a microscope*

*An early eighteenth-century Hebrew scholar's attempt to explain aspects of human physiology in mechanical and architectural terms*

*Francesco Redi, who made the earliest attempt to refute the theory of spontaneous generation experimentally. He showed that while putrefying meat in unsealed jars spawned fly maggots, in sealed jars it did not*

physiology as those of Malpighi and Harvey were isolated advances, linked primarily with medicine. They did not readily fit into any theoretical framework, except to show that the human body and the bodies of animals were pieces of machinery subject to mechanical laws. But even this conclusion was not easily absorbed, for it clashed with the religious and philosophical views of the Middle Ages. These were based on even earlier philosophies: the Greek idea of the separation of body, mind and soul has left its imprint on Western thought to this day.

## The generation of life

Some of the most significant of the early work in biology was concerned with the 'spontaneous generation' of life. Educated men believed that mice were generated from hay and flies from meat. To them this was after all more credible than that stones fell from the sky. Despite the outstanding experiments of Francesco Redi on fly maggots in Italy about 1680, it was not finally cleared up until the time of Louis Pasteur in the mid-nineteenth century. Pasteur's careful work finally showed that spontaneous generation does not occur, and that all living creatures are only produced from others of their kind. This of course still leaves open the problem of the first origin of life, and whether it could arise on planets other than our own, and this we shall take up in later chapters.

The dominating figure of biology in the eighteenth century was the Swedish taxonomist Carolus Linnaeus (1707-78). Single handed, he set himself the task of cataloguing, describing, arranging and naming the whole world of organized nature. This was a phase of data collection, and it was not till Charles Darwin (1809-82), with his Theory of Natural Selection, that any great attempt was made at explanation, if we except a few studies of physiology in animals and plants. Darwin's work was based on the labours of Linnaeus and the taxonomists; he showed that the observed diversity of life could best be explained by assuming that all living creatures had evolved over long periods of time from one primitive life form. We think now that this first living creature probably arose over 3,000-million years ago. Recognizable microfossils have been found dating from about 3,200-million years ago, while at the beginning of the Cambrian period, about 600 million years ago, all the major groups of animals were present.

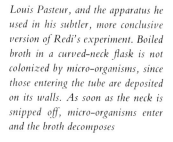

*Louis Pasteur, and the apparatus he used in his subtler, more conclusive version of Redi's experiment. Boiled broth in a curved-neck flask is not colonized by micro-organisms, since those entering the tube are deposited on its walls. As soon as the neck is snipped off, micro-organisms enter and the broth decomposes*

Darwin's great contribution was to provide a reason for this evolution: *the fittest survived best*. Out of the many variants that constantly occur in a stock, those best fitted to their environment are perpetuated by natural selection. This principle is now seen to operate in many fields. It is not even restricted to biology, for we see echoes of it in subjects as diverse as astronomy and nuclear physics, where the idea of progression from less stable to more stable states under the pressure of the environment is a counterpart of the more complex concepts of biological evolution. The more stable sorts of star, the more stable isotopes, accumulate, and the less stable diminish.

The other great biological discovery of the nineteenth century was the elucidation about 1860 of the basic principles of genetics by Gregor Mendel (1822-84). Mendel's work was curiously neglected until the turn of the century, when it was rediscovered and rapidly developed. It was shown that the determinants of genetic potential were carried on separate entities (as they were then thought of) which were called genes, and the basic behaviour of genes was found to apply to animals and plants of all kinds. Only in bacteria was this in doubt, but we shall see later that it was from studies on this problem group that our biggest recent advance, the physico-chemical mechanism of the genes, was to be elucidated. Two major steps were soon taken. The American geneticist T. H. Morgan showed that the genes were carried on the chromosomes—rod-shaped bodies that largely compose the nuclei of cells, and which consist of deoxyribonucleic acid and protein. By means of elegant experiments on the fruit fly *Drosophila*, Morgan and his colleagues demonstrated that the genes were strung along the chromosomes like beads on a necklace, each gene having a fixed position on a particular chromosome. The way in which genes behave was elegantly explained by the behaviour of the chromosomes during the production of the eggs and sperms and their subsequent fusion during fertilization to give the fertilized cell that initiates the next generation.

*The Bohemian monk Gregor Mendel, father of modern genetics*

About this time, too, the statistical work of R. A. Fisher first offered a clear explanation of the way evolution works. He showed that selection of the fittest can be understood in terms of the change in the relative frequencies of different forms of the same genes (alleles) in an interbreeding population. Rare genes arise by mutation (sudden structural changes) and give rise to the variant gene alleles. Under the pressure of selection—due to the environment—the alleles that make the organisms best fitted to the environment increase in the population, and these replace the older, less suitable, alleles.

*Above: isolated chromosomes of the midge* Chironomus tentans. *Their characteristically banded appearance has enabled investigators to locate positions of specific genes. The stunted wings of the fruit flies (left) are due to a mutant gene*

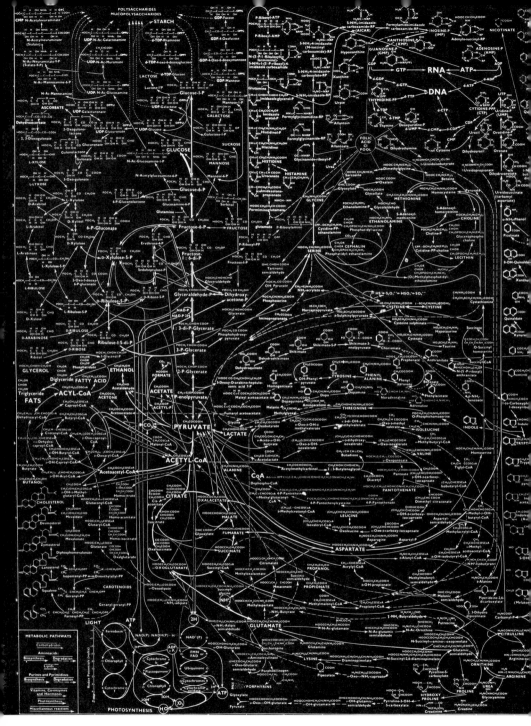

METABOLIC PATHWAYS

PHOTOSYNTHESIS

## The chemical unity of life

The further extension of genetic knowledge had to wait for several decades, but in the meantime biochemists were uncovering the chemical structure of living organisms. It became clear that all organisms are based upon proteins and nucleic acid. Proteins, besides forming much of the mechanical structure of the cells (e.g. structural proteins, like tendon) are also the chemical catalysts that control the biochemical reactions of cells. These proteins are called enzymes, and as a rule one enzyme carries out only one specific chemical reaction. Certain kinds of enzyme and certain chemical reactions were found to recur in the most diverse forms of life. Many workers contributed to this, but the theory of the biochemical unity of life was due in large part to the work and writings of A. J. Kluyver. We now have a detailed map of most of the metabolic pathways of living creatures, differing only in small details in man, bacteria and other organisms. This map, although not well known outside biochemistry, is one of the wonders of the age, a veritable *mappa mundi mirabile*. It shows, as scarcely anything else does so well, the chemical unity of life.

The structure of proteins was unravelled slowly. Proteins are composed in essence of long chains of amino acids of which 20 different sorts are normally found, and all 20 occur in every kind of organism. The work of Frederick Sanger at Cambridge showed that the arrangement of the amino acids along the protein chain is fixed for a given protein. Sanger's brilliant detective work in piecing together small fragments from chemical analyses into the complete chain also explained why proteins are specific to the job each performs. The unique sequence of the amino acids gives the protein molecule its unique properties; it makes it fold up in such a way that it forms a three-dimensional structure which fits the shape of the chemical on which it has to act; this also brings to the right position those amino acids (or other chemical groups fixed to the protein) that do the actual job of catalysing the reaction in question. The twenty

*Opposite: a remarkable 'map' of the metabolic pathways involved in the cell chemistry of all living creatures. It suggests something of the complexity of living matter*

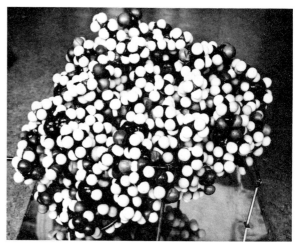

*Two models of the protein molecule myoglobin, the one emphasizing the convoluted path of the amino-acid chain and the other showing the positions of individual atoms*

different amino acids allow enormous variety of protein structure, and thus protein molecules can be 'tailor made' for the most diverse chemical jobs. A protein only 10 amino-acid molecules long can have $10^{20}$ alternative structures (because there are twenty alternatives at each position of the chain), and this figure is 100 million million million. Most proteins consist of chains of several hundred units, so the number of possible proteins is vast, far more than the number of protons and electrons in the observable universe. It is thus easy to see that proteins can be 'tailor made' not only for particular chemical reactions, but also for the rather different conditions in which they may have to work in the cells of the millions of species of living creatures that exist upon the earth.

But what of the nucleic acids? Nucleic acids are not the only other class of chemical compound found in all living creatures. The amino acids themselves, and many of the intermediate chemical agents involved in the energy-yielding reactions of cells, are also universally present. (Consider, for example, the active molecule ATP (adenosine triphosphate), which is always found as a powerful transmitter of chemical energy.) But they are of fairly simple structure and their number is relatively small. It was at first thought that nucleic acids were also small molecules of limited variety, but this view was altered by the discovery of O. T. Avery that the chemical molecule that determines the genes in bacteria is pure deoxyribonucleic acid (DNA). Deoxyribonucleic acid is also a molecule in the shape of a chain, a very long chain (indeed of indefinite length). It is composed of units each consisting of sugar phosphate (deoxyribose phosphate) and a 'side group'. The side group is always one of the four bases adenine, thymine, guanine and cytosine. A complete unit (base, sugar and phosphate) is called a nucleotide, so that another way of describing DNA is as a chain of nucleotides.

Since deoxyribonucleic acid is found in chromosomes, it was clear that the genes might be composed of DNA. The enormous variety of DNA molecules could store the necessary information to code for the proteins and carry the genetic blueprint of the organism. Just as given proteins have a fixed sequence of amino acids, so it was realized that DNA sequences could spell out messages in an alphabet of four letters (A, T, G, C for adenine, thymine, guanine and cytosine), which could determine the same messages in a different form—the protein sequence in an alphabet of twenty letters (one for each of the 20 sorts of amino acid). It was also clear that changes in the sequence of DNA could account for gene mutations,

*The essential elements of DNA: alternate phosphate and sugar units, the sugar linked to one of four possible side groups*

Thymine     Cytosine     Adenine     Guanine

*The DNA molecule, which may comprise as many as 300 million atoms, exists in the cell as a double helix. Fully extended it would measure nearly a millimetre in length. Right: an atomic model of a small section of the helix*

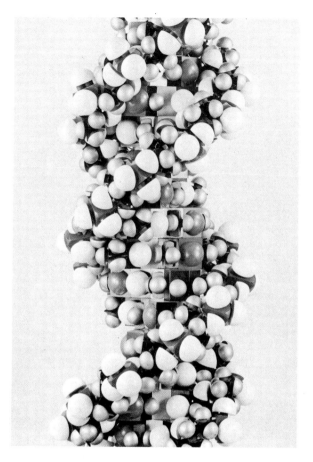

*The twin sugar-phosphate chains forming the DNA helix are united by hydrogen bonds between their side groups. Thymine can only pair with adenine and cytosine only with guanine*

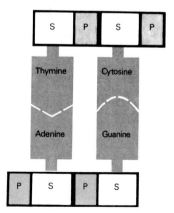

because the new sequence would give rise to an altered protein. Two problems remained. The first was how the gene message in DNA could be copied exactly, so that a descendant would have the same genetic message as its parent. The second was what was the DNA code that specified the amino acids.

### The genetic code

The first question was answered by J. D. Watson and F. H. C. Crick at Cambridge. They showed that the DNA molecule is double, consisting of two chains twisted round each other in a double helix (like a two-stranded rope). They postulated that only certain pairs

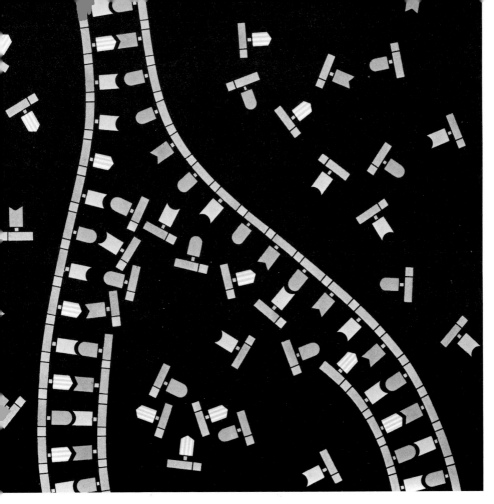

of the four bases could bond to one another along the helix, that is, adenine to thymine, and guanine to cytosine. Thus the only possible combinations are: A-T, T-A, G-C or C-G. This would mean that if a helix unwound to give two single strands, and two new strands were built up on the single ones, these new strands could only have the complimentary sequences of the original strands. The four strands would now form two double helices. The DNA would thus have given rise to an exact copy of itself. This theory has now been confirmed by several lines of evidence, and is rightly considered to be an historic step forward in biological knowledge.

*Replication of the DNA molecule: the chains separate and each, combining with available free units, becomes the template of a new chain and a new helix*

| | | | | | | | |
|---|---|---|---|---|---|---|---|
| UUU | Phe | UCU | Ser | UAU | Tyr | UGU | Cys |
| UUC | Phe | UCC | Ser | UAC | Tyr | UGC | Cys |
| UUA | Leu | UCA | Ser | UAA | stop | UGA | stop |
| UUG | Leu | UCG | Ser | UAG | stop | UGG | Trp |
| | | | | | | | |
| CUU | Leu | CCU | Pro | CAU | His | CGU | Arg |
| CUC | Leu | CCC | Pro | CAC | His | CGC | Arg |
| CUA | Leu | CCA | Pro | CAA | Gln | CGA | Arg |
| CUG | Leu | CCG | Pro | CAG | Gln | CGG | Arg |
| | | | | | | | |
| AUU | Ile | ACU | Thr | AAU | Asn | AGU | Ser |
| AUC | Ile | ACC | Thr | AAC | Asn | AGC | Ser |
| AUA | Ile | ACA | Thr | AAA | Lys | AGA | Arg |
| AUG | Met | ACG | Thr | AAG | Lys | AGG | Arg |
| | | | | | | | |
| GUU | Val | GCU | Ala | GAU | Asp | GGU | Gly |
| GUC | Val | GCC | Ala | GAC | Asp | GGC | Gly |
| GUA | Val | GCA | Ala | GAA | Glu | GGA | Gly |
| GUG | Val | GCG | Ala | GAG | Glu | GGG | Gly |

*The genetic code. The four side groups of DNA may be regarded as a four-letter alphabet, yielding a possible 64 three-letter 'words'. Of these words 61 code for the 20 amino acids; the remaining three are 'full stops' signifying the end of a protein chain.*

| | | |
|---|---|---|
| U—uracil (thymine) | Cys—cysteine | Met—methionine |
| C—cytosine | Gln—glutamine | Phe—phenylalanine |
| A—adenine | Glu—glutamic acid | Pro—proline |
| G—guanine | Gly—glycine | Ser—serine |
| Ala—alanine | His—histidine | Thr—threonine |
| Arg—arginine | Ile—isoleucine | Trp—tryptophan |
| Asn—asparagine | Leu—leucine | Tyr—tyrosine |
| Asp—aspartic acid | Lys—lysine | Val—valine |

The second question was posed in particular by Crick's speculation on how an alphabet of four letters could give rise to an alphabet of 20 amino acids. Quite recently this has been solved by M. Nirenberg and others. Three bases code for one amino acid. The genetic code, which applies universally to living creatures, is shown in the table opposite. It will be noted that thymine (T) is not listed, because during the translation process thymine is replaced by uracil (U). T and U are therefore equivalent.

The way in which DNA is translated into protein is somewhat involved. The DNA, with its A-T-G-C code, is translated into another form of nucleic acid—ribonucleic acid (RNA)—in which thymine is replaced by uracil as mentioned, but in which the other letters are unchanged. The ribonucleic acid is called 'messenger RNA', and consists of a fairly short chain (some hundreds of links long) which codes for a single protein. This A-U-G-C message is translated into the protein message on intracellular structures called ribosomes, where the protein is assembled from amino acids, each carried by a small RNA carrier. It should also be noted that the genes can no longer be thought of as beads, but rather as lengths of DNA, arranged end to end along the chromosomes.

*Proteins are manufactured in regions of the cell cytoplasm called 'ribosomes' (large dark areas). The code for a protein is copied from a section of DNA by a short chain of 'messenger RNA' (1), which passes into the cytoplasm. There free amino acids (2) are captured by small molecules of 'transfer RNA' (3) bearing their code. These are brought to the ribosomes (4) moving along the messenger RNA, and are incorporated in the growing protein chain*

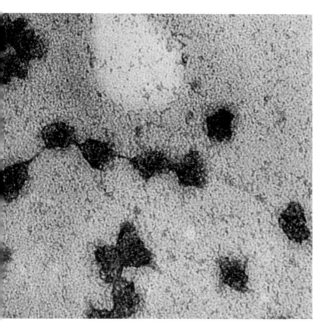

## The variety of living things

The amazing variety of living forms is a constant source of wonder to biologists. We are familiar enough with various vertebrates—mammals, birds, fish—yet there are more species of insects than of all other animals put together, and some of them are extremely bizarre. Every niche—in biology, a particular combination of life-supporting conditions—has been occupied by some or other form of life, and almost every imaginable activity has been realized in some living creature. There are plants that count up to two and trap insects, fungi that catch nematode worms, lichens that are composite organisms, consisting of a fungus and an alga living in partnership, an arrangement known as symbiosis. Even the lowly slime moulds and bacteria show some extraordinary forms.

The variety of structures, shapes and sizes of living things is characteristic of life on our planet, and one would expect this to be true wherever life exists. Behaviour is similarly diverse. Some organisms are sessile, others move: some get their energy from sunlight, others are predators, herbivores or scavengers. Their chemical capabilities are equally varied, as we shall see below. But living things also differ in another way from the non-living (besides the three usual

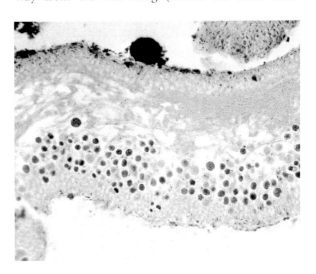

*Section through a lichen revealing a symbiotic partnership: small photosynthetic algae lie within a protective network of fungus cells*

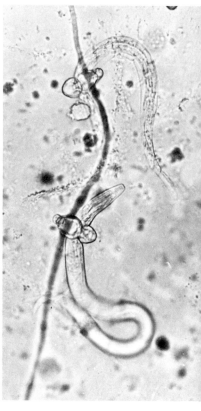

distinctions of feeding, growing and reproducing). They interact with one another as well as with their environment in an intricate and far-reaching way. A rock has little influence on another rock. A bird or a tree or moss, however, affects its neighbours profoundly, and may often influence events over large distances. Living creatures together form 'ecological' systems that are as complicated as the internal systems within an organism, and perhaps more so. At any rate, we are only just beginning to understand how to investigate ecological problems. (In a very broad sense this is part of the new field of systems analysis, which investigates the factors at work in extremely complicated interrelationships, such as hold among living systems themselves.)

*Further unexpected forms of life. Above left: a carnivorous plant—the sundew—traps and digests a fly. Above right: the fungus* Arthrobotrys oligospora *captures and kills a nematode worm*

Life occurs on Earth under a wide range of conditions, very much wider than we may perhaps imagine, if we think only of familiar animals and plants. Some form of life is found almost everywhere on the earth's surface. Even in the Arctic and Antarctic, on the most barren rocks, there are usually some lichens and mosses, and in the shelter of these, and under stones, one can usually find insects and other small animals. The same is true of the highest mountain peaks. In these environments the temperatures may be fairly high in the sun during the day, but at night it is intensely cold.

But the smallest organisms, micro-organisms like bacteria, have even more remarkable powers. Some of these are extraordinarily hardy and can survive and even multiply under very adverse conditions.

*Edelweiss on an alpine crag—an example of a plant adapted to withstand unusually low temperatures*

## Conditions for survival

Biological multiplication is possible over virtually the whole range of temperatures between the freezing point and boiling point of water. At the lower temperature living processes become very sluggish, and for most organisms cease for practical purposes at freezing point. There are, however, some marine animals and micro-organisms that can grow at a few degrees below freezing point. For example, there are fish that live in seawater below 0°C whose blood is supercooled, and which therefore run the risk of freezing solid should they touch a piece of ice. Some algae still swim actively in salt pools at −15°C.

Organisms in very cold situations must be protected in some way from frost if it is not to damage their tissues, and many of them have developed their own 'antifreeze'. A substance like glycerol (commonly known as glycerine) acts in the cells in the same way as 'antifreeze' does in a car radiator, and sugars have a similar effect. This allows these creatures to withstand cold periods and return to activity when it becomes

*Life is also possible in unusually hot environments: certain algae and bacteria can live in thermal springs at temperatures of up to 85°C*

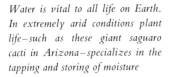

*Water is vital to all life on Earth. In extremely arid conditions plant life—such as these giant saguaro cacti in Arizona—specializes in the tapping and storing of moisture*

warmer. The temperature under snow is often not as low as one might expect, however, and various animals and plants can live under a snow field—indeed some plants can produce enough heat to melt the snow in their immediate vicinity (e.g. snowdrops and the alpine *Soldanella*).

At the other extreme are creatures living in hot springs such as those at Yellowstone National Park in the United States. These springs often show striking colours—red, yellow or green. Some colours are due to inorganic pigments, but many result from heavy growths of micro-organisms, principally the blue-green algae. These, together with a few bacteria, are capable of living in water between 70° and 80°C. Although dormant stages of micro-organisms can resist much higher temperatures, the upper limit for active growth seems to be about 95°C, at least in nature, for C. E. ZoBell reports one example of growth at a higher temperature than this in the laboratory. He found that a heat-loving bacterium, which preferred about 80°C under normal conditions, continued growing at 104°C under a pressure of 1,000 atmospheres (one atmosphere is about 14·7 lb. per square inch, the air pressure at sea level on the earth). The effect of pressure in this case is to simulate lower temperatures. It has been suggested that at these high temperatures there is a constant race between the destruction of proteins by heat and their rapid replacement by the synthetic mechanism of the cell, and it is certainly true that heat-loving bacteria can multiply very rapidly. But it is also likely that the proteins of these heat-loving organisms are more resistant to heat damage than those of most organisms.

All living things need water, but some organisms obtain it in strange ways. Desert plants may send roots down hundreds of feet in search of water or, alternatively, store enough water from the brief rains to last them for the rest of the year. Some moth and beetle larvae that live on very dry materials, like wood or wool, obtain their water from their food, by retaining the water formed during its digestion. Clothes-moth grubs are a familiar example. A few desert rodents do the same, and never need to drink. Yet some source of water is essential, and in extreme deserts, such as the Atacama in South America, life becomes very scanty. Micro-organisms in general need rather moist conditions, and for this reason drying is a good way of preserving things; in desert cemeteries, for example, bodies may not decay for millennia. However, it has recently been pointed out that very dry soils may contain considerable amounts of salts that can absorb moisture from the air and give rise from time to time to a film of saturated solution around the soil particles; some micro-organisms may then be able to grow in this film.

Lichens are particularly well adapted to extremes of temperature and moisture. They are found on barren rocks in deserts and in the arctic regions. They grow mainly during short periods of warmth and damp and become dormant again with the return of

*Most temperature resistant of all plant groups, lichens are equally at home in the Sahara and the Arctic*

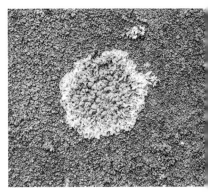

more extreme conditions. Lichens, as we have seen, are symbionts, a hardy fungus providing shelter for an alga. They need only traces of inorganic nutrients, which may be provided by airborne dust or the rocks on which they grow. The algal elements synthesize food from water and atmospheric carbon dioxide using the energy of sunlight captured by photosynthesis.

The next requirement is a source of energy. Most living creatures obtain their energy either from sunlight, as in green plants, or else by eating other organisms. There are some surprising sources of food. Some insect larvae actually live and thrive in pools of crude oil. The most curious examples, though, are chemoautotrophic bacteria—bacteria which obtain their energy from the oxidation of inorganic chemicals. *Thiobacillus* oxidizes sulphur to sulphuric acid, for example, and other bacteria oxidize hydrogen to water, or nitrite to nitrate. *Ferrobacillus* is particularly interesting, because the energy obtained by oxidizing ferrous iron to ferric is very small, and this bacterium must have developed some ingenious way of boosting this low-grade energy to the high levels needed to make essential chemicals like ATP.

Although one naturally thinks, next, of air as a requirement of life, even this is not needed by many creatures, notably anaerobic bacteria. A familiar example is the common brewer's yeast, which converts sugar to alcohol and carbon dioxide in the absence of air in the deep brewing vats.

Micro-organisms can tolerate wide extremes of acids and alkalis. *Thiobacillus* will grow in solutions containing 3 per cent of sulphuric acid, not far below the concentration of acid in a car battery. Alkaline mud flats provide highly alkaline conditions where micro-organisms will grow. Strong salt solutions, too, are no hindrance to some; bacteria and brine shrimps live in saturated brine in the Great Salt Lake in Utah. Recently an icy pool containing 33 per cent calcium chloride yet supporting living micro-organisms was found in Antarctica.

*Yeast, a micro-organism which in deep brewing vats converts sugar to alcohol and carbon dioxide, does so in the complete absence of air*

High pressures, again, have little effect on many living creatures, for some bacteria and higher animals live on the deepest sea bottoms at pressures of over 7 tons per square inch. Experimentally, some bacteria have survived pressures of nearly 60 tons per square inch. Low pressures have no obvious effect on micro-organisms provided liquid water is still present. Spores of bacteria and fungi are readily collected at high altitudes in our atmosphere.

In a liquid medium the effects of gravity are probably offset almost entirely by pressure. We therefore have no particular reason to suppose that weightlessness, as in space flight, will have any profound effect on living creatures. The successful manned space flights to date

*The effects of short periods of weight-lessness do not seem to be significantly harmful to living organisms, but little is yet known about the effects of prolonged weightlessness*

*Opposite: the core of a nuclear reactor viewed from the top of the insulating pool – hardly an environ-ment hospitable to life; yet bacteria capable of withstanding intense radiation have been found in such pools*

show that there are no short-term effects of great importance. Nevertheless there have been a few reports that weightlessness interferes in some poorly understood way with cell division and differentiation, but much more work is needed before this can be evaluated. On our present knowledge we should at least expect micro-organisms to flourish in a weightless condition.

Resistance to X-rays and certain other radiations is still poorly understood. Higher organisms are rather sensitive to such radiation: a dose of a few hundred roentgens will kill most mammals and plants (a roentgen is a unit of radiation dose; a typical hospital X-ray delivers rather less than one roentgen). Yet bacteria and blue-green algae grow in the radioactive cooling ponds where fuel cans from nuclear reactors are stored, and even in the cooling water of the reactors

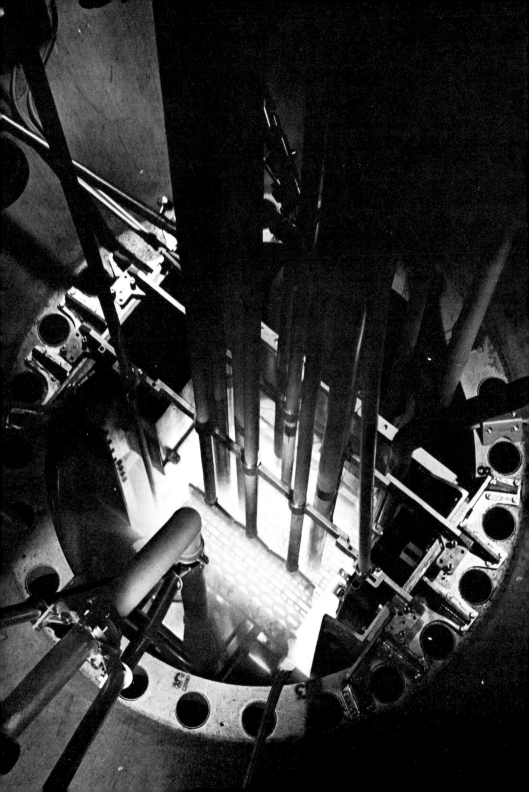

themselves. They can resist about a million roentgens. Recent work shows that their resistance is due largely to certain protective chemicals within their cells. Relatively slight shielding can confer a great deal of protection. We ourselves are well shielded by the thick blanket of the atmosphere. X-rays and gamma rays from the sun, and charged particles, are absorbed by the sheer mass of the atmosphere, which is equivalent to several feet of lead. Ultra-violet light is largely absorbed by the ozone layer high in our atmosphere. Only a little ultra-violet and a few cosmic rays reach the surface of the earth. But even if the other radiation did (and it may have been much more intense in the early days of our planet), it might not have very much influence on micro-organisms. True, higher land animals and land plants could not survive, but the radiation-resisting lower forms would survive, together with marine creatures that were protected by a few feet of water. Again, in outer space, or on the surface of the moon, unprotected micro-organisms would be killed fairly quickly, but a small amount of shielding would protect them substantially.

## Longevity and dormancy

Man is short-lived by some standards, long-lived by others. Trees appear to be the most long-lived of plants (bristle-cone pines growing in California have been found with as many as 4,500 annual growth rings) and among animals certain tortoises. In 1966 there died Chief Tu'imalila, a giant tortoise given by Captain Cook to the King of Tonga in 1777, and accorded by the people of Tonga the honours due to a chief. But, in order to make significant journeys into space, long life would be less important than the ability to remain dormant. Dormancy—the state in which organisms survive without growth or metabolism—is therefore of central interest.

Dormant organisms are often relatively resistant to both high and low temperatures. Sometimes this is related to desiccation, for most micro-organisms are much more heat resistant when dry. Bacterial spores

are the most heat resistant of all forms of life. When quite dry they require almost charring temperatures to kill them; when moist they can survive boiling in water for some hours, though they are killed within an hour in moist steam under pressure at 120°C. Recently it has been reported that a very small proportion, but some nevertheless, of bacterial spores survived the brief but intense heat of passing through the flame of a rocket motor: about one in 100,000 survived flame temperatures of about 3,000°C lasting a few thousandths of a second. Bacterial spores are also often

*Bristle-cone pines, grotesquely stunted in appearance, may seem doomed, yet of all multi-cellular living things they come closest to immortality. Some specimens still thriving are estimated to be more than 4,000 years old*

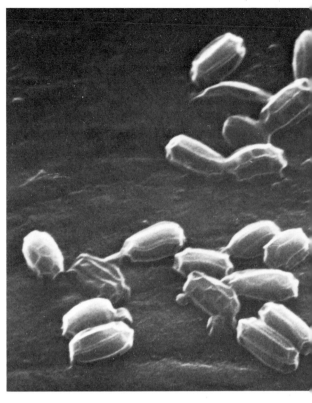

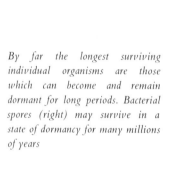

*By far the longest surviving individual organisms are those which can become and remain dormant for long periods. Bacterial spores (right) may survive in a state of dormancy for many millions of years*

resistant to chemical disinfectants and are usually unaffected by a high vacuum. Most micro-organisms, too, can survive intense cold, such as that of liquid helium (−269°C), and in fact they are now often stored in a deep-freeze to preserve them for laboratory purposes.

Few higher organisms are so robust, but H. E. Hinton of Bristol University has reported on the larva of a small fly called *Polypedilum vanderplanki* which lives in transient pools in Africa. When the pools dry up the larvae survive in the dried mud. Provided they are quite dry they can survive temperatures as high as that of boiling water, or as low as that of liquid air (−190°C). (They are killed, however, by quite dilute alcohol.) The larvae revive within a few minutes when

*The notion of dormancy exploited in science fiction: 'cybermen' revive and clamber out of their survival cells after an extended period of total inactivity*

*So hardy are bacterial spores that some, it has recently been discovered, are able to pass unscathed through the intensely hot flame of a rocket motor*

moistened. One unexplained point of great interest is how the proteins of the larva are protected against heat, whether by some protective chemical, or because they have a peculiar heat resisting structure.

It has long been known that some seeds can lie dormant for many years. Seeds of the lotus have germinated after being buried for a thousand years in a peat bog. Recently S. Ødum, a Danish scientist, studied the seeds lying dormant in undisturbed soil under various archaeological sites (hearths, walls, etc.), and reported that seeds of the white goosefoot and corn spurrey had survived for over 1,700 years. There has also been a report of germination of seeds of the lupin *Lupinus alpinus* after several thousand years of burial in permanently frozen arctic silt.

Extreme cold may promote the survival of dormant organisms. The seed which gave rise to this arctic lupin, for example, recently germinated after being buried for 10,000 years in frozen silt in Canada

Bacterial spores also survive for long periods. I have looked at some old soil samples on the roots of pressed plants to see how long soil bacteria survive. Soil collected in 1640 still had many viable spores, and it appears that a very few would survive for as long as a thousand years. A few older materials that I examined —from a burial in a peat bog about 1,000 years ago, and from a desert, about 2,500 years old—did not show any living bacteria, but these materials had been preserved under conditions that are not ideal for dormancy. Several workers have looked for bacteria surviving in coal or beds of rock salt, but their findings have usually been very inconclusive, and it is extremely difficult to exclude recent contamination. Similarly we have no reliable information of bacteria in the frozen mammoths in Siberia, though bacteria have survived in frozen materials from Captain Scott's antarctic expedition at the turn of this century.

There are difficulties in reconciling very long survival with what we know of survival curves. Moreover, we believe that chemical decomposition over long periods would be enough to kill the micro-organisms. Organic compounds decompose—though very slowly—at ordinary temperatures. P. H. Abelson has shown that in fossils some of the amino acids in proteins have disappeared. In time one would expect both structural materials and the unstable substances essential to metabolism to be lost.

However, at very low temperatures chemical break-down would be negligible, and bacteria could probably survive for millions or even billions of years at temperatures close to absolute zero (though this is based on extrapolation from higher temperatures, and may not be entirely accurate). In outer space, tem-peratures are very low, so that in theory life could survive in meteorites or dust particles (the high vacuum of space would be unlikely to kill micro-organisms).

At very low temperatures a second factor is import-ant, however, and this is the accumulated dose of radiation. Much of this may come from solar or stellar bombardment, but there is some also from natural radioactivity of the material composing the carrier particles, and indeed also from the matter that composes the organisms themselves. All living creatures contain potassium (K), and the isotope K-40 is weakly radioactive, so that some internal radioactive disinte-grations occur. Nevertheless the lethal effect of radiation is probably unimportant for periods of less than a few million years. These points are of special relevance in considering in later chapters whether life could have spread through space to our world.

In the last chapter we explored many of the major features of living organisms as we know them on our planet; in this chapter we shall ask what generalizations we can make about the living state wherever it might occur in the universe. We shall clearly be on uncertain ground in extrapolating from our knowledge of terrestrial biology. But since it would seem likely that the variety of life on the earth is narrower than the variety of life in the universe at large, it seems most appropriate to aim for wider concepts and definitions than those we usually employ.

## The concept of living

It is usually said that living creatures show four unique properties: they grow, feed, react to their environment and reproduce. We shall examine each in turn.

Living organisms grow from within, and in a dynamic manner. Molecules of the body are removed, as well as added, during growth. This allows reorganization of the entire body structure as required. For example, in our own bodies there is a slow turnover of the atoms of which our bodies are composed. Some atoms are rapidly exchanged for new ones provided by our food, drink and air (for example, those in energy stores such as fat). Structural material like bone is turned over more slowly, but substantially all of this is replaced by new atoms over the course of years. The adage that we become new men every seven years

*Opposite: a man and his wife dwarfed by the monolithic trunks of the giant Californian sequoia—largest known living thing*

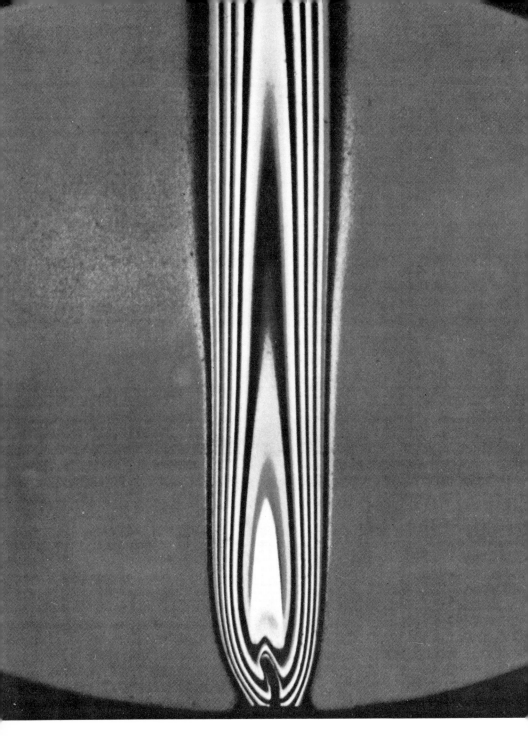

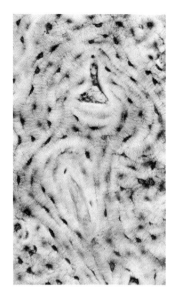

therefore has a grain of truth. Yet we remain the same persons.

Most non-living things that 'grow' do so by simple accretion from outside. Crystals do this. But there are some non-living things which show the self-regeneration characteristic of living creatures. A candle flame, for example, is ever changing, yet always the same. So we see that there are exceptions to the rule about the manner of growth of living things.

Living organisms take in food for two purposes, to supply energy and to provide the material for growth and replacement. But, again, a candle flame does the same. Nor can we use the kind of food as a criterion of life, for we have seen that there are bacteria which have the simplest sorts of nutrition; green plants 'feed' on sunlight and carbon dioxide, while we prefer steaks and Château Lafite.

The way living things react to changes in their environment is also not easy to define in rigid terms. Some, like non-motile bacteria, show little evidence

*Non-living structures, if they grow at all, tend to do so by accretion from without, as in the case of the growth spiral on a silicon carbide crystal (above left). Living structures, such as bone tissue (above), grow and renew themselves from within. The candle flame (opposite), able to regenerate and maintain itself, mimics a living entity, but lacks its complexity*

of reacting at all, although a study of their biochemistry would show that they adapt their metabolism to changes in their food supplies and other features of their environment. Higher organisms have elaborate sense organs, like ears and eyes, and while we can postulate these as necessary for intelligent life forms, we cannot say they are necessary to life itself. A flame, too, reacts to its environment—to a draught for example—and both flame and living organism tend to adjust themselves in such a way as to preserve their metabolism in the most steady condition possible. It is not for nothing that the poets have spoken of 'living' flames. Flames can even reproduce in a sense, but here our analogy breaks down, because the daughter flames in a spreading fire are unlike their parents in physical form, although similar enough in their chemical nature and general construction. It is thus to reproduction that we look for a sharper distinction between the living and the non-living.

The reproduction of living organisms has two characteristics that are rare in non-living systems, or only faintly developed. Living organisms produce very faithful copies of themselves, but where variations do arise these variations tend to be perpetuated. Moreover, the variations are subject to selection by the environment, so that over numerous generations the organisms evolve to give new forms. This, of course, is the core of Darwin's theory of evolution, and it is difficult to envisage any life forms that would not be subject to it in some way. Natural selection would appear to be inevitable if organisms produce close copies of themselves; inevitably these copies would sometimes be defective, and inevitably those more fitted to the environment would be favoured. It is true that the environment might be so constant, and the creatures so well suited to it, that no possible change in their genetic make-up could give any advantage. But it is very improbable that such a situation would endure for any length of time.

The possibility that living organisms might not reproduce is an idea very foreign to our way of

*Evolutionary variations on a theme : the effect of natural selection on the shape of ammonoid shells at various stages in their evolutionary history*

c.290m.yrs.

c.260m.yrs.

thinking. Even science fiction writers have seldom tried to grapple with it. It has a flavour of the metaphysical about it, for such beings would be virtually immortal, and their origin would pose questions like the origin of angels. Self-creating but non-reproducing living creatures would indeed be strange. Fred Hoyle's 'Black Cloud' is one of the few attempts to get to grips with this, for Hoyle suggests that certain sorts of aggregation of cosmic dust might begin to show organized electrical properties and then become self-organizing, so that they become increasingly complex, 'alive' and self-aware—fatherless and motherless, yet sentient. (We may note in passing that science fiction writers seldom attempt to justify the physiology of their imaginary creatures; this example of Hoyle's, and some of H. G. Wells's, are noteworthy exceptions.)

We might take this idea further. If life can arise at all, then it could arise repeatedly, if suitable conditions persist for long enough. Given these, we could imagine creatures that came into being yet neither died nor reproduced, but modified themselves in response to the environment. Evolution and selection would then operate internally on their constitution, rather than on a succession of descendant organisms.

The nearest analogy is perhaps that of soils. Soils do not reproduce in the ordinary sense, yet they are complex organized systems (that include in this case many separate living things). They arise wherever rock is exposed to the air, and they respond to changes in the environment; they show variety, but only within certain limits, and persist indefinitely. If we imagine living things with properties like these, such living things may not be wholly improbable. Their existence could mean that life of the sort regularly arises from non-living matter. It seems likely, too, that such creatures could not compete with one another (just as soils do not compete), otherwise they would blend harmoniously with their fellows and lose their own individuality.

Yet another theme is taken up in many forms by science fiction writers. Robots need not reproduce;

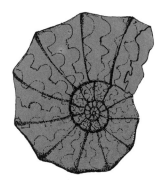

c.180m.yrs.

c.140m.yrs.

c.110m.yrs.

59

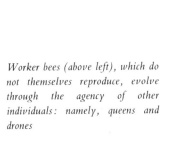

*Worker bees (above left), which do not themselves reproduce, evolve through the agency of other individuals: namely, queens and drones*

their production is not necessarily reproduction. Men make them, and (in various tales) the robots live, are mortal (they wear out), or are immortal (they are self-repairing). The robots only evolve through their makers' plans. Other stories make the robots truly reproducing, and they would then be capable of evolution. Some terrestrial organisms show certain parallels. Worker bees, wasps, ants and termites do not reproduce. Selection acts on the workers and the whole nest, and the plans of the workers evolve through the medium of the queens and drones. One could draw a similar parallel from the cells composing our bodies, for it is only the specialized sex cells that give rise to the next generation.

The traditional concept of life, therefore, may be too narrow for our purpose. It is true that this is a matter of the definition of life, which we will not pursue here, but a broader concept, of 'para-life', if you will, may serve us better. We should try to break away from the four properties of growth, feeding, reaction and reproduction.

Is there anything that underlies these four? Perhaps there is a clue in the way we speak of living *organisms*. They are *highly organized*, and perhaps this is indeed their essence.

## Organism and organization

What, then is organization? What sets it apart from other similarly vague concepts? Organization is perhaps best viewed as 'complex inter-relatedness'. Inter-relatedness means that the various parts of the complex pattern affect one another in action and reaction. A book is complex; it only resembles an organism in that passages in one paragraph or chapter refer to others elsewhere. A dictionary or thesaurus shows more organization, for every entry refers to others. A telephone directory shows less, for although it is equally elaborate, there is little cross-reference between its entries. We can readily understand, too, how the different components of soil—the rock, sand and clay, the humus and the multitude of creatures living in it and on it—all interact with one another. The young subjects of cybernetics and systems analysis are primarily concerned with inter-relations and interaction, but they have not yet given us any very profound insights into life—except to emphasize how subtly involved it is, how numerous are the interacting factors, and how difficult it will be for us to reach any deep understanding of the whole. Nevertheless, we have learned something. Living systems are in metastable equilibrium with their environments, which in simple terms means two things: first, that they react to changes in the environment in such a way as to preserve their internal integrity, or more generally, they try to minimize the disturbing effects of their surroundings; and second, that they can do this only within rather narrow limits, for if these are exceeded, living systems lose their stability—they die and decay. This self-correcting power is seen from the amoeba that avoids an irritant particle to the human society that adjusts itself to the political and economic pressures of its day, while death overtakes all systems that are subject to stresses they cannot tolerate. We should be on strong ground in assuming that meta-stable equilibrium is a characteristic of any system we would call living. Indeed this is perhaps why a dictionary, however involved, would be excluded, for it cannot adapt itself if you tear out some pages: its inter-relatedness is static, not dynamic.

Dynamic inter-relatedness, then, is indispensable to organization, but so is complexity. A candle flame shows quite elaborate inter-relatedness: the wick, melted wax, flame and fumes react on each other. It shows self-correction: if the wick is snipped short, the flame diminishes, and the molten wax is not all vaporized as it runs up the wick, so that the wick is burnt away less quickly. The wick thus reverts to its usual size. Yet a candle flame is not very complex; it has too few parts distinguishable by form or function to be considered alive.

Living organisms are characteristically very complex. The metabolic map illustrated on page 30 shows this clearly, for it represents a complexity that is found in essentials in all living creatures that have ever been investigated biochemically. Their complexity is shown too by their diversity. This diversity is two-fold. Living creatures differ from one another, so that we see the myriads of different forms that have evolved over the ages. They also show diversity within themselves; they have cells, organs and parts, and even the simplest bacteria show elaborate internal structure.

Complexity is measured by the information required to specify the system, and we can try to measure this for living organisms. First, we need to distinguish between complexity that is repetitious and that which is not. Repetitious information is found in living organisms, it is true. Much of the organism is made from a limited number of building blocks. For example, cellulose is made of a repeating chain of glucose molecules. This allows great flexibility in

*Cellulose, an example of repetitious complexity found in living organisms, consists of long, simple chains of repeating glucose units. Each of the many interwoven threads shown here comprises many hundreds of such molecular chains*

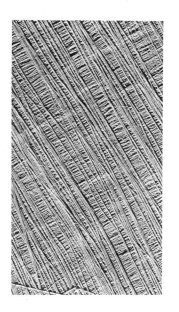

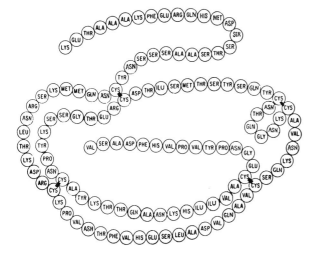

*Proteins exemplify non-repetitious complexity. Shown here is a molecule of the enzyme bovine ribonuclease consisting of a long chain of amino acids arranged in a precise sequence and cross-linked at definite points. Any structural alteration is liable to render the enzyme ineffective*

construction. The indefinite length of the cellulose chain allows it to be just as long as is needed in any given set of circumstances. This is particularly important for the materials forming the mechanical fabric of cells. It is an advantage, too, to have a limited variety of building blocks. Terrestrial organisms seem to manage with thirty or forty of these. Repetition also increases safety. Two eyes are safer than one. A centipede can afford to lose a leg or two. The same principle is believed to operate in our brains, for we believe that several duplicate nerve-cell circuits partici- pate in any thought process. The end result then depends on a sort of majority verdict. This reduces the chance of disastrous error due to failure of a single component, and it is this principle which is used in parallel-logic electronics, concerned with producing ultra-safe machines.

In contrast to repetitious or redundant information, we have the exact length and precise sequence of an enzyme, a protein which is to act like a very precise 'machine tool' rather than like a wall of identical bricks. Here we cannot compress the information, because there is virtually no repetition. The bulk of the infor- mation required to build an organism is probably needed for the numerous enzymes.

## The information content of organisms

We must first be clear about what is meant by information content. Very simply, information has to be viewed as a *difference* between two states of knowledge. Let us take an example. If one wished to send a cable to South America giving the current price of various commodities—wheat, barley, maize, and so on—one could send a message in full—leaving out only obviously redundant words—such as 'Wheat X shillings per bushel'. Much more compact would be to arrange a code with our correspondent, W for wheat, B for barley and so on, and to agree that prices were always in shillings per bushel. Then we would simply send 'W X'. In everyday life we would say that both cables had the same content of information, but in the scientific sense this is not so. The short message carries little information because a lot of information is already possessed by the receiver, who knows the meaning of the code. With some oversimplification we can say that information content depends on the length of the message (as long as redundancy has been removed).

In biology we are not sure just what information should be implied. However, the information must be that for a *working* organism. A random arrangement of chemical molecules in a solution requires as much information to define it as their arrangement in a living cell. But while any of the random arrangements in a solution are much alike for any practical purpose, and none are 'alive', to make a living cell we must have certain very particular arrangements, and only these will do. Evolution has been called a mechanism for generating improbabilities, and this is so in the sense that the genetic messages in living creatures are highly specific: it is very improbable that they would arise by chance, if the molecules joined up at random.

We shall take, as the basis for discussion, the information content in an alphabet of four letters, like the genetic code, but whatever our basis the contrast between the living and the non-living is very great. A crystal such as sodium chloride (common salt) consists

*The crystalline structure of common salt (left) can be exhaustively described in a few words by referring to the positions of sodium and chlorine atoms in a simple three-dimensional lattice (below). Contrast the information content of such a description with that which a description of the metabolic map on page 30 would entail*

of a repeating pattern of sodium and chlorine atoms in a three-dimensional array. If we take the atoms, their number, and their spacings as the implied information, then a small sodium chloride crystal could be described as a 'cubic-lattice structure with spacing of $2 \cdot 82 \times 10^{-8}$ cm., consisting of alternate sodium and chlorine atoms, and containing $10^{20}$ atoms'. This goes on to a few lines of print.

But consider even the simplest of living creatures, a bacterium. It embodies the metabolic map to which we have referred. It contains perhaps 10,000 different enzymes, each with a precise sequence of 100 to 200 amino acids. Some variation in the sequence may perhaps be allowed, but not very much. It also has a complex internal structure. All of this cannot possibly be expressed in a few lines of text. How much information, then, does the DNA of a bacterium contain, and is it enough to code for all the proteins involved?

A convenient measure of information is the 'binary digit', or 'bit'. This is a single piece of information which is simply either 'Yes' or 'No', usually written, for convenience, as 1 or 0. As it happens, these two possible 'bits' can code for four alternatives. We can thus represent the four nucleotide bases of the genetic code as: adenine $= 11$; thymine $= 10$; guanine $= 01$; cytosine $= 00$.

We could then write a message of 1s and 0s which would carry the genetic message of the bacterium. Now the length of the DNA of a bacterium is about 10-million nucleotides, or 20-million bits. If it were printed out in full in the letters A, T, G, C, it would fill several thick books. The number of different messages that could be written would be around $10^{6,000,000}$, far more than the number of electrons and protons in the observable universe (which is only about $10^{125}$). This number of 20-million bits would comfortably code for about 20,000 different proteins, so we have accounted for over 10,000 enzymes with a bit to spare.

One might ask whether much of this DNA message was repetitious. Could not the code for one enzyme be repeated a hundred times? There are several indications that in bacteria there is very little repetition. In higher organisms we are in more doubt; perhaps quite a large fraction of the DNA consists of repetitions, maybe nine-tenths, and there is some evidence that this is so. Nevertheless, there seems to be a general agreement between complexity (as measured by the number of kinds of tissue cell—muscle, skin, liver, etc.) and the content of DNA in the chromosomes. The simplest organisms have a shorter genetic message than the more complex ones, which need in addition the extra information for controlling harmoniously the functions of the different organs and tissues. Yet there are some species with an unexpectedly high DNA content, and the most plausible explanation seems to be that these are species in which there is a great deal of repetitious DNA.

Bacteria are the simplest living creatures that can multiply independently. Viruses do not qualify, since they can only reproduce within the cell of a suitable host, where they direct the cells's mechanism towards their own purposes, and divert it from its normal

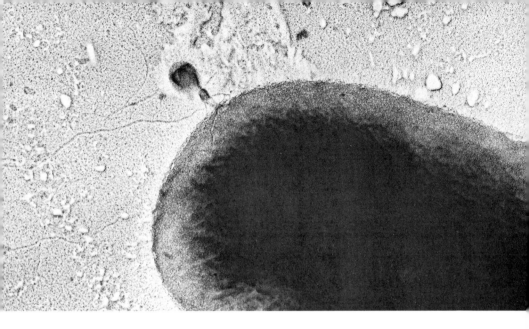

functions. Whether viruses are living is a question that has long exercised the philosophical biologist. It depends, of course, on just what you mean by 'living'. But in an everyday sense viruses are on the borderline between the living and non-living. Evidence is accumulating that some of them are by origin parts of the normal cell which have acquired a degree of autonomy. This allows them to survive transmission from host to host, but they do not contain the full genetic message for copying themselves. They therefore rely on the copying mechanism of the host, as well as on building blocks produced by the host cells. Significantly, their DNA content is well below that for bacteria.

Is there any theoretical minimum for the information content of an organism that is able to copy itself? The mathematician John von Neumann calculated that a machine which copied itself would not need more than 32,000 parts, and would need a blueprint of only 150,000 bits. He was concerned simply to prove by logic that such an entity was possible (as indeed all forms of life asserted), and not to find the minimum number of bits. His theory, too, is of uncertain application to biology, because the minimum depends on the

*Viruses, unable to grow or reproduce outside a host cell, are on the borderline between living and non-living. Above: a bacteriophage, a virus that attacks and infects bacteria, injects its contents into the cytoplasm of a bacillus*

*Isolated viruses, of a kind causing poliomyelitis in man, may assume an almost crystalline aspect*

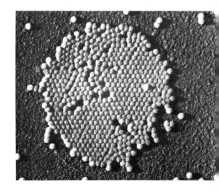

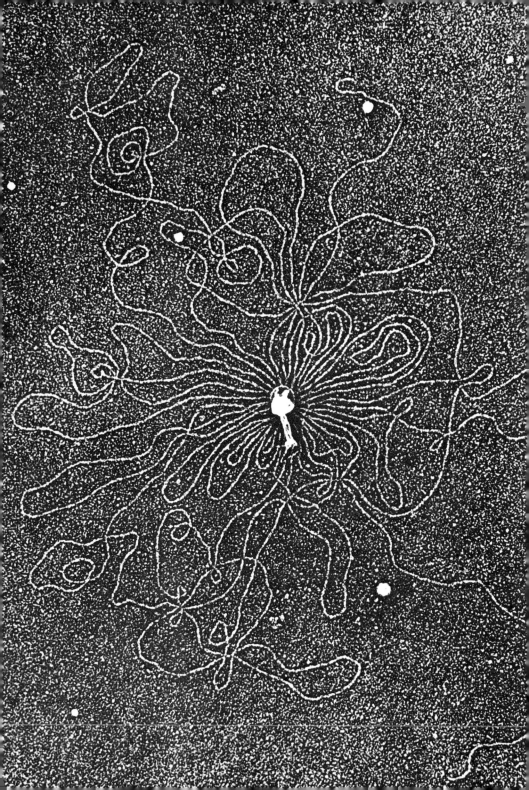

sort of building blocks that are supplied. If we supplied adenine, thymine, guanine and cytosine nucleotides, plus energy, and if no catalyst were needed, we could imagine that a DNA molecule, however short, might copy itself indefinitely. In theory the minimum number of nucleotide pairs would be 1, or rather 2, because, as we shall see, some uncoupling mechanism is also needed, and a second pair would have to control this. Yet if we require the DNA molecule also to carry instructions for producing these building blocks from carbon dioxide, ammonia, water and phosphate it is clear that it would have to be longer. The nearer to the right form and variety that the blocks supplied are, the less information the organism needs to store in order to manufacture them. The metabolic map probably represents a substantial fraction of the 20-million bits in the DNA of a bacterium, so the bacteria are presumably near the lower practicable limit of information content for a free-living organism. With viruses the position is rather different. Viruses do not have, so far as we know, the metabolic apparatus of living organisms. They have some mechanism for assembling building blocks provided by the cells in which they grow, and probably also part of the mechanism involved in the replication process. These can perhaps be coded in small messages, so there may be no lower limit that we can set at present on the nucleic-acid content of viruses.

A very neat demonstration of many of these points has been given by L. S. Penrose of London University, who uses simple models to illustrate the principles involved. The 'organisms' are represented by cut-out figures, which assemble themselves into new configurations, replicate themselves, and so on. Part of the 'information' is provided by the complex building blocks themselves. One may make a rough estimate of the information they contain by assuming that this is the number of 'bits' that would be required to instruct a computer-controlled fretsaw machine to cut them out, and this comes out at several hundred bits. If we use three-dimensional instead of plane figures, it seems

*Opposite: a remarkable electron micrograph of a burst bacteriophage. The length of the extruded DNA molecule reflects the vast information content of even the simplest of living organisms*

likely that the corresponding information content of such models would be around $10^5$ 'bits'.

Penrose's models are very imaginative. They illustrate, incidentally, an important requirement for replication: some sort of information transmission *within* the organism. Thus in order to join two pieces together there must be some sort of hook. Yet when this pair replicates, the hook must somehow be undone again, so information must be transferred to it at the right time. It would seem very worthwhile for chemists to study Penrose-type systems, simple though they may be. Could organic molecules be designed with the equivalent of his hooks and catches?

## The physical medium of life

We have been discussing generalities about organisms and organizations, but little has been said about their physical form. What restrictions are there on this? The microbiologist C. J. Perrett has argued that a membrane or border is essential to a living cell. This is a point that has received little thought—even less the question of whether it must be composed of matter. It seems to us inconceivable that non-material organisms could exist—for example, organisms composed of radiation patterns, which would need something to keep them from diluting themselves by radiation into space. Nevertheless, the fact that all the atoms in an individual can (at least in theory) be changed during growth without loss of the organism's identity, shows that it is the pattern of organization that determines life, rather than simply its material form. The fact that living organisms require a source of energy is also important. Energy flow is needed to *do* things, and in its absence, life can at most be suspended, as in dormant spores. Active life requires an energy source (a source of high energy) and an energy sink (a source of low energy). Our own energy sources are the food we eat and the air we breathe; our energy sink is our surroundings, into which the energy from our food runs away mainly as heat-loss. In some sense, therefore, the sources and sinks must be separated from one another, with the

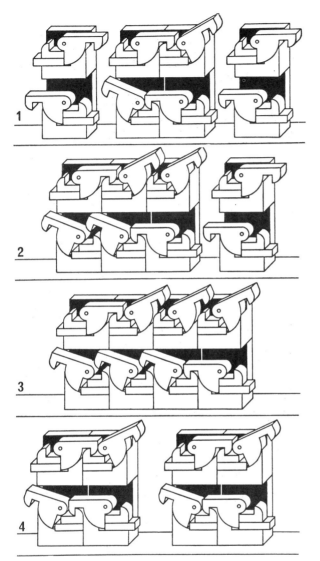

*One of Penrose's ingenious self-replicating models. A two-unit 'organism' in the presence of free units (1) links with a third unit (2). Linking with a fourth unit, however, causes the original pair to separate (3) leaving a pair of identical two-unit 'organisms' (4)*

organism between them. Barriers must therefore exist between the organism and the sinks. But what form the barriers might take is a larger question. They would have to be energy barriers of course, but whether they could be gravitational, electrical, magnetic, or operate on some other principle, does not seem to have been widely discussed.

71

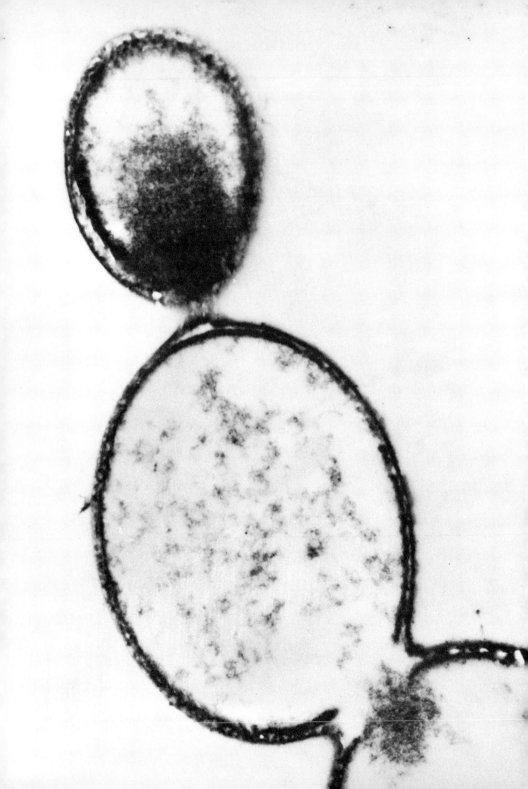

# THE ORIGIN OF TERRESTRIAL LIFE                    4

How did life on Earth originate? We have seen that early naturalists accepted that life was generated spontaneously from organic materials like hay, dung or flour. But the researches of men like Redi, Tyndall and Pasteur showed that life is only produced from pre-existing life. Darwin's theory of evolution implied that all living things had descended from one primordial life form, and our new knowledge of the unity of biochemistry and of the genetic code has triumphantly vindicated this belief. The origin of terrestrial life was therefore pushed back into the dim past, some 3,000-million years ago, and it was generally felt that nothing could usefully be said about it. Darwin was one of the first to speculate in the modern fashion, and he noted the difficulty of demonstrating the formation of life by any experiment. In 1871 Darwin wrote:

> It is often said that all the conditions for the first production of a living organism are now present, which could ever have been present. But if (and oh! what a big if) we could conceive in some warm little pond, with all sorts of ammonia and phosphoric salts, light, heat, electricity, etc., present, that a protein compound was chemically formed ready to undergo still more complex changes, at the present day such matter would be instantly devoured or absorbed, which would not have been the case before living creatures were formed.

*If artificially synthesized amino-acid chains are cooled in solution, microscopic spheres, or droplets, are seen to separate. Those opposite possess a double enclosing layer which suggests a primitive cell membrane*

## The panspermia hypothesis

Very little was written on the origin of life until the publication of the Russian scientist A. I. Oparin's book *Origin of Life* in 1923, later expanded in a second edition in 1936. Such theories as were propounded claimed little attention. The usual view is that life originated on the earth from non-living matter after it had cooled to a moderate temperature. Only two fundamental alternatives were proposed. One of these, which can scarcely claim much support, was by W. Preyer: the fiery liquid mass of the earth was already an organism, and as it cooled the protoplasm of living creatures separated from the dead parts which formed the rocks. The other alternative was most convincingly advocated by the great Swedish chemist S. Arrhenius, who called it the panspermia hypothesis. Arrhenius suggested that life arrived on the earth in the form of germs such as the spores of micro-organisms, and that these had been carried across the depths of space from elsewhere in the universe, propelled by the weak but continued pressure that is exerted by light rays on minute particles.

Both these theories, of course, simply push back the origin of life by one more step, but the panspermia hypothesis is at least open to some critical assessment, as well as having intrinsic interest. Arrhenius showed theoretically that light pressure would be able to move small particles over very great distances. He also showed that violent volcanic eruptions might carry such particles into the stratosphere, from whence they might be driven into interplanetary space. He calculated that particles might reach the nearest stars, the Alpha-Centauri group, in some thousands of years. However, larger particles would tend to fall by gravitation into the sun. Arrhenius thought that the earth may have received its first life from spores transported from planets elsewhere in the universe, and believed that the extreme cold of outer space would not cause them to perish during their journey, nor would the transient heating as they entered the earth's atmosphere.

Indeed there are many plausible arguments in the

*The Russian scientist A. I. Oparin, pioneer of research into the origin of life on Earth*

panspermia theory, and the astronomer Carl Sagan of Harvard University has re-examined many of these points. A spore of diameter 0·4 to 1·2 microns (1 micron = 1/1000th of a millimetre) is of the right size to be pushed out from the solar system by the sun's light. It would reach the orbit of Mars within a few weeks, and the nearest stars, as Arrhenius said, in some tens of thousands of years. A spore rather larger than this would be attracted to the sun (because for it the gravitational pull would be greater than the radiation pressure), and might hit the earth *en route*. Spores of quite a variety of sizes could be expelled from other planetary systems, depending on the mass and brightness of their suns. In outer space bacterial spores might survive for thousands or even millions of years, provided they had some protection against ionizing radiation, such as might be afforded by adsorbed dust particles.

The most serious difficulty is that of space itself. Space is so vast that even if cubic miles of spores were liberated in it they would become so dispersed in the enormous volumes of space that there would be little chance that a planet such as the earth would ever capture any of them. In any case the great majority would be burned up by the stars or in hot gas clouds. Such calculations are necessarily speculative, but they would make it seem very improbable that life could have reached the earth from elsewhere. There may, however, be some possibility of transport from the earth to other planets in the solar system. Thomas Gold has suggested an alternative method of transport: the earth may once have been visited by an expedition from some advanced civilization elsewhere in our galaxy, and microbes may have been left behind!

## Abiogenesis

If we return to the more plausible view that life arose on the earth itself, we may distinguish three main views: (1) that life arose as a supernatural event, which is therefore outside the scope of scientific discussion; (2) that life originated from common chemical

*Stanley Miller and the apparatus with which in 1953 he performed his famous 'life-synthesis' experiment, producing a variety of simple organic compounds from an inorganic mixture resembling the earth's primordial atmosphere*

reactions by a slow and inevitable evolutionary process; and (3) that life originated as a very improbable event, but one which would occur naturally whenever suitable conditions were present for long enough.

It was J. B. S. Haldane who pointed out the importance of distinguishing between (2) and (3), although they grade into each other. If the time required for the improbable event under (3) is many times greater than the lifetime of a planet, this would make life a very rare event indeed, although not supernatural.

Before the writings of the Russian scientist A. I. Oparin, in the 1920s and 1930s, it was usually thought that life arose in an atmosphere containing free oxygen. Oparin argued that the first life originated when the earth's atmosphere lacked free oxygen, and instead consisted of reduced compounds like methane and ammonia, together with nitrogen and water vapour. About this time, too, Haldane pointed out that the metabolic pathway of sugars in most organisms is much the same for the steps which do not require oxygen, so that it looks as if the primitive pathway was anaerobic. He, too, favoured an atmosphere without oxygen. Oparin suggested that organic compounds, mainly hydrocarbons, arose from the

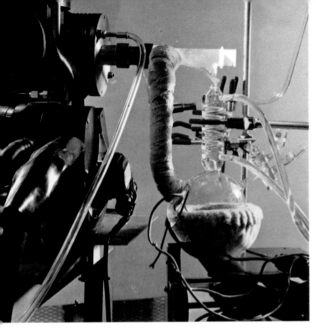

*Apparatus used in a recent primordial synthesis experiment at the Chemical Evolution Division of the Ames Research Center in California. Since Miller's experiment still more complex organic compounds – including nucleotides and ATP – have also been synthesized*

reaction between metallic carbides (in the hot rocks) and water, and that these hydrocarbons reacted further with water and ammonia to give a whole range of the simpler organic compounds, like amino acids. This is thought to have been going on some 3,500- to 4,000-million years ago, after the earth had cooled enough to provide pools of liquid water. Clearly there are two main things we would like to understand: the way in which the simpler organic compounds formed, and the way these became organized to give the first living thing. This early period of the earth's history was the phase of 'abiogenesis'—the production of complex organic compounds without life.

There was a good deal of discussion, led by the American chemists M. Calvin, H. C. Urey and their associates, on how organic compounds like amino acids might arise, because experiments in mixtures containing free oxygen seemed to yield only small traces of the sort of substances that were needed. Then in 1953 at Chicago University S. L. Miller performed a famous experiment by following Urey's suggestion that an anaerobic (oxygen-free) atmosphere was necessary. Miller passed an electric discharge through a mixture of methane ($CH_4$), ammonia ($NH_3$)

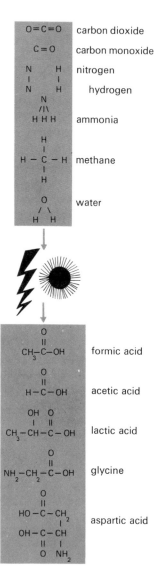

*The probable first steps towards life: examples of organic compounds likely to have resulted from the effects of lightning and ultra-violet light on the earth's primordial atmosphere*

O=C=O    carbon dioxide

C=O    carbon monoxide

nitrogen

hydrogen

ammonia

methane

water

formic acid

acetic acid

lactic acid

glycine

aspartic acid

and hydrogen ($H_2$) in a flask containing water, and in order to make the non-volatile compounds accumulate, continuously distilled the water through a sealed circuit. After a week the water was deep red, and contained, in addition to simple acids like acetic acid and formic acid, at least two amino acids. Furthermore, there was evidence for the presence of hydrogen cyanide (HCN), which is known to be an active compound capable of producing many more-complex derivatives. Since then it has been shown by Calvin, Urey, C. Ponnamperuma, J. Oro and others that there are many agents that can produce organic compounds (some of them quite complex) under similar conditions. In the early days of the earth there must have been plenty of energy for such syntheses, from lightning and ultraviolet light.

It has also been suggested that meteorites or comets may have contributed large amounts of complex carbon compounds over the course of millions of years. Organic materials were first found in meteorites by the chemist Berzelius in 1834. The spectra of comets indicate complex carbon compounds, and they may have struck the earth from time to time. Indeed it has been suggested that the great Tunguska meteorite of 1906 in Siberia was the head of a comet.

### The dawn of life

We now have no difficulty in seeing how amino acids, nucleotides, sugars and the like could arise. It is more of a problem to understand how they could become aggregated so as to form the first living cell. Indeed, we should not really talk here about living cells, because the earliest living forms may not have been cells at all. This stage in the process has been called the phase of eobiology (dawn life) or protobiology (first life), and the organisms accordingly eobionts or protobionts. Oparin pointed out that if various large molecules like proteins are in solution together they sometimes aggregate into droplets called coacervates. In these several kinds of molecule are held together within a sort of skin formed by certain of the

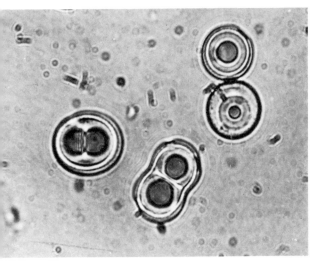

*Amino acids heated in the laboratory form long molecular chains and, on cooling, protenoid droplets (left). Such structures may conceivably have been the forerunners of organized proteins and, ultimately, of cells (see illustration on page 72)*

larger molecules. These droplets may absorb water and swell, and also divide. He thought that life could have arisen from coacervates which gradually acquired additional properties as they floated around in the 'thin soup' that formed the early oceans. It has been shown by W. S. Fox that one can, by heating amino-acid mixtures, produce protein-like chains which will form globules that behave in much the same way as coacervates, and there have been claims that these globules may also possess weak enzymic activity. It seems likely, too, that nucleic-acid chains can also be formed from polyphosphate esters, which in turn can arise from ATP formed abiogenically, that is, in the absence of life.

J. D. Bernal thinks it more probable that the simple organic molecules were absorbed on to particles of clay, and that this may have greatly accelerated the rates of chemical reactions which they underwent. It is of course possible that droplets containing both coacervates and clays may have formed, and it has been argued that the most likely site for the origin of eobionts would have been in shallow pools that periodically dried up, for this would increase the concentration of organic compounds, and facilitate a wider range of chemical reactions than in the oceans.

We cannot tell whether life originated once (in one particular coacervate or clay particle) or whether the same steps were being repeated all through the early seas and pools. But at some stage the power to multiply arose, probably in the form of a nucleic-acid chain similar to those we know today, and presumably using something very like the present genetic code. One would then expect the most successful eobionts to multiply and over-run any other incipient organisms.

At this stage metabolism would depend on energy provided by the organic molecules which the eobionts decomposed anaerobically. These eobionts, between 3,000- and 4,000-million years ago, might have been something like bacteria, though lacking many features of present-day bacteria. They would possess a cell membrane, within which would be enzymes and nucleic acids. They would use as building blocks the preformed amino acids and nucleotides which were dissolved in the seas. As time went on, and some building blocks became scarce, these eobionts would have to develop processes for making them from other organic molecules, or else would have to adapt to use substitutes. Natural selection would act to perpetuate the most successful forms, and evolution would begin.

These eobionts presumably possessed three major attributes. They had self-copying nucleic acids, they had enzymes, and they had a membrane. In what order did these arise? This is, of course, just speculation, but it seems most likely that they arose in the order given above. The copying mechanism came first, because it is this that defines the continuity of structure which is implied when we say it came first. The nucleic acids then acquired the power to absorb amino acids and form them into proteins, while the membrane was then evolved to stabilize the whole. Recently, however, it has been plausibly suggested that the first nucleic acids formed on a protein chain (which may not have been an enzyme in the usual sense), so that one could envisage an early period when the self-copying molecules were a complex of nucleic acid and protein.

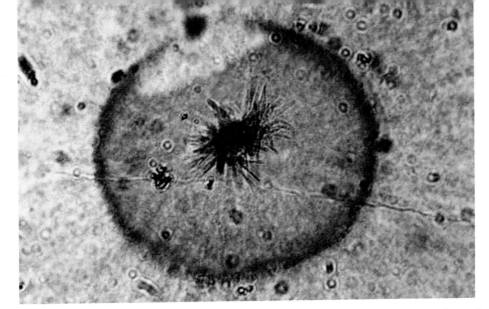

There is much discussion at present on how the genetic code originated. The third base of the nucleotide triplets listed in the table on page 36, usually does not alter the amino acid that is coded for, and this suggests that in eobionts the first two letters only were read, and that the third was utilized later to distinguish certain pairs of amino acids. The eobionts may well have had a limited repertoire of amino acids. But it seems rather unlikely that the original code consisted of twin letters only, because of the difficulty of changing from a twin to a triplet code. But whether the first code contained only adenine and thymine, or only guanine and cytosine, no one knows.

Among these bacteria-like creatures (which by now would be using up the available 'food' that had been produced abiogenically) some would develop chemical mechanisms for using energy from sunlight. These forms would give rise to the blue-green algae (which, like the bacteria, are very simply constructed). The blue-green algae had probably evolved by 3,000-million years ago, and they would have begun the production of atmospheric oxygen through photosynthesis. By now the abiogenic soup would have been exhausted, and sunlight would be the main source of energy.

*One of the earliest known forms of life on Earth, this minute fossil alga is more than 2,000 million years old. Also present in the same rock – Gunflint chert from Canada – were other microfossils and minute traces of the breakdown products of chlorophyll, suggesting that photosynthesis was already occurring. Similar traces have also been detected in 3,000-million-year-old rocks in South Africa*

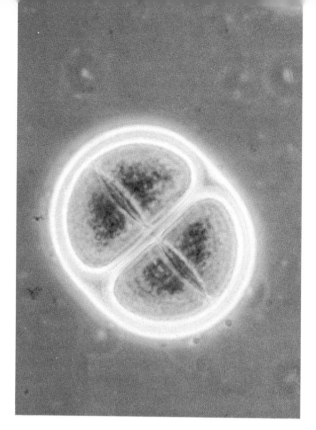

*Procaryotes are primitive cells whose nuclear material is not contained within a nuclear membrane. Eucaryotes—which form the basis of all plant and animal life—are more advanced cells, possessing a well-defined nuclear membrane. Right: the four daughter cells of the blue-green alga* chroococcus turgidus—*a procaryote. Opposite: the nucleus and part of the cytoplasm of an eucaryote, here an animal cell*

## The evolution of higher organisms

Around 2,000-million years ago a major change in organization took place. In certain organisms the cell nucleus developed a well-defined membrane of its own, and from these organisms arose all the higher forms known as eucaryotes ('having a good nucleus') which include protozoa, animals, fungi and green plants. The bacteria and blue-green algae are procaryotes; they lack a nuclear membrane and on several grounds are thought to be most like the earliest organisms. Near this time, too, certain new structures arose in the cytoplasm, notably mitochondria, important sites of energy production. We think it was somewhat later (c. 1,500-million years ago) that certain eucaryotes acquired green chloroplasts, the structures in which photosynthesis in higher plants is carried on. Mitochondria and chloroplasts have features suggesting that they may originally have been bacteria and blue-

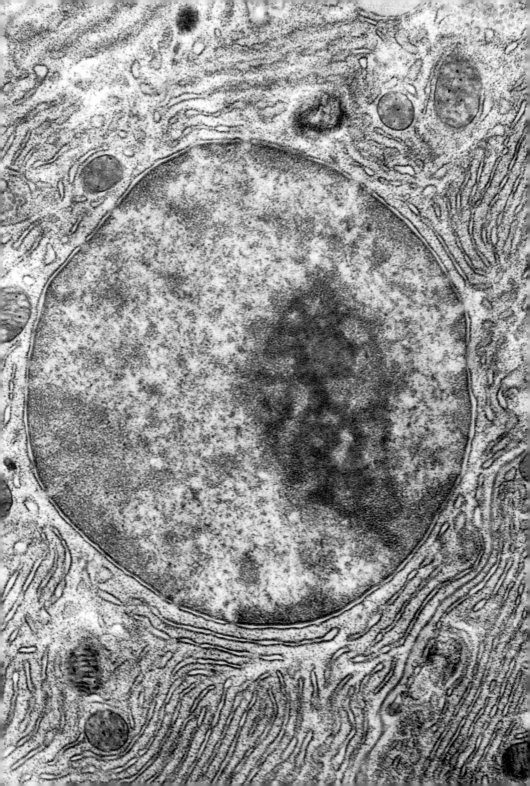

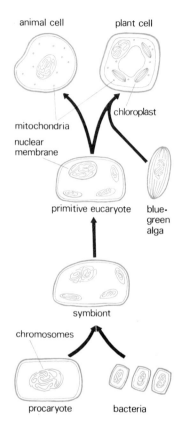

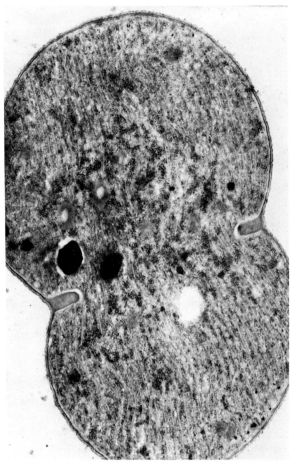

green algae respectively. If so, we ourselves, with all higher organisms are symbiotic, while green plants represent a triple partnership formed from stocks of the original eucaryote, bacteria and blue-green algae. If this hypothesis is true then evidently two symbiotic unions occurred. First a bacterium became symbiotic within the forerunner of the early eucaryote, and later a blue-green algae associated with certain of these to form a trio. From the trio came the green plants. The animals and presumably the fungi came from the earlier double organism.

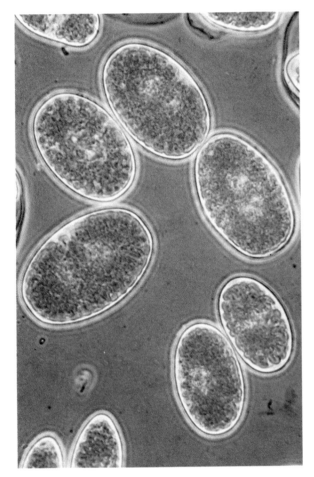

It is possible that symbiosis has played a crucial part in the evolution of plant and animal cells. Mitochondria—the cell's energy centres—may originally have been bacteria which became incorporated in the cytoplasm of an early procaryote. Chloroplasts—the photosynthetic agents in plant cells—were probably once blue-green algae which entered into a symbiotic union with a primitive eucaryote. Opposite: a blue-green alga (about to divide). It has no mitochondria or chloroplasts, but possesses a photosynthetic apparatus which ramifies throughout the whole cell. Below: a chloroplast. Left: the green alga Glaucocystis nostochinearum—a symbiont. Lacking chloroplasts, it is served instead by blue-green algae

By now methane and ammonia would be scanty. The increasing oxygen in the atmosphere would allow animals to rise, feeding on the green plants or their breakdown products. The first animal fossils are dated somewhere around 800-million years ago, and with the dawn of the Cambrian period, 600-million years ago, we find all the major groups of invertebrate animals. Land plants and animals arose rather later, about 300-million years ago, and well before this time the atmospheric oxygen would have produced the ozone layer high in the stratosphere, which absorbs

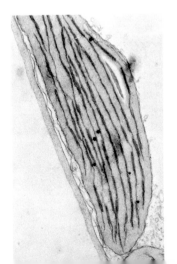

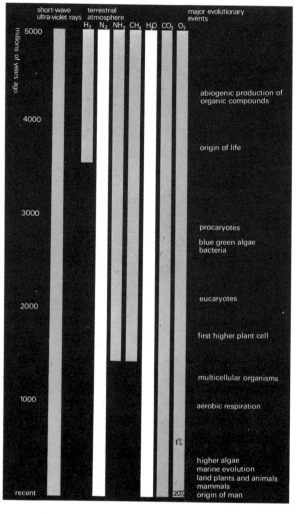

short-wave
ultra-violet rays

terrestrial
atmosphere

H₂  N₂  NH₃  CH₄  H₂O  CO₂  O₂

major evolutionary
events

millions of years ago

5000

4000

3000

2000

1000

recent

1%

20%

abiogenic production of
organic compounds

origin of life

procaryotes

blue green algae
bacteria

eucaryotes

first higher plant cell

multicellular organisms

aerobic respiration

higher algae
marine evolution
land plants and animals
mammals
origin of man

*Some of the more important
changes in the composition of the
earth's atmosphere related to the
estimated dates of various major
evolutionary events*

most of the harmful ultraviolet light in sunlight (this
would not have mattered to the marine creatures,
because water also absorbs ultraviolet light). Mammals
arose about 200-million years ago; by contrast mankind today is scarcely a million years old.

In later chapters we shall discuss whether it is likely
that life is common throughout the universe. Much
therefore depends on our answer to the question: Does
life originate quickly and regularly in a suitable

environment, or *does it require an event so improbable* that the chance of its arising in any reasonable time is extremely small? Opinions have been much divided on this, and except for opinions there is very little to go on. Many who work on abiogenesis, such as Calvin and Urey, believe that life will arise regularly on suitable planets in the times available. Others, such as the biologists H. Blum and G. G. Simpson, believe that this is very unlikely. Haldane was not too sanguine. He pointed out that two things were critical in the argument—the minimum size of a nucleic acid that could replicate, and the number of alternatives of that minimum size that would replicate. If only one specific message in a nucleic acid of 250 links was suitable, then the chances that this one would arise by accident would be vanishingly small, even over vast periods of time. We have seen in the previous chapter that the minimum length of a nucleic acid may perhaps be quite small. But we still do not know what the chances are that the right organic molecules will come together in the right pattern for long enough to start the copying process. We probably do not yet know how to calculate them, and it is a fair guess that any estimates at present would be subject to such tremendous uncertainty that the exercise would be of very doubtful value. We shall instead assume that the chances of life arising on other worlds are good enough to justify our inquiries, but leave it at that.

# LIFE BEYOND THE EARTH

In this chapter we shall review the evidence for life on other planets. The table overleaf shows the planets of our own solar system, with the properties of most interest to us. Upon the distance from the sun depends to a large extent the temperature of the planet, and this distance is therefore one important measure of its suitability for life. The mass is important too, for if it is too small the planet's gravitational field would be too weak to prevent any atmosphere from escaping. A planet, we believe, must retain its atmosphere for thousands of millions of years if life is to arise upon it. The lower limit of mass for this is around that of Mercury—but it also depends on the temperature. A hot planet will lose its atmosphere more quickly than a cold one. What is important is the temperature at the upper limit of the atmosphere, for it is here that the hottest and lightest gas molecules are lost by shooting off into space. Large planets, like Jupiter, have such strong gravitational fields that they lose very little of even the lightest of gases, hydrogen. Planets of medium size lose a good deal of hydrogen, as the earth itself must have done, and this tends to leave behind the heavier gases, like nitrogen and carbon dioxide. A good deal may have been lost and replaced by volcanic gases from their interiors. Probably the giant planets, like Jupiter, have kept almost all their original atmospheres, however. The composition of planetary atmospheres is estimated mainly by analysing their

*Space biologists are now planning to establish by on-the-spot experiments whether life exists elsewhere in our solar system. That such an undertaking is no longer an empty dream is due above all to the current development of efficient space transport systems. Opposite: the awesome Saturn rocket a few seconds after 'lift-off'. Its celestial destination —the moon—is visible as a narrow crescent through the launching smoke*

| Planet | Average distance from sun (millions of miles) | Length of year (Earth time) | Length of day (Earth time) | Diameter (miles) | Mass (Earth:1) | Density (Water:1) | Escape velocity (miles/s) | Average surface temp. °C |
|---|---|---|---|---|---|---|---|---|
| MERCURY | 36 | 88d | 59d | 3,100 | 0·05 | 3·8 | 2·2 | 400/-180 |
| VENUS | 67 | 225d | 250d | 7,700 | 0·82 | 5·1 | 6·3 | 400 |
| EARTH | 93 | 365d | 23h 56m | 7,927 | 1·00 | 5·5 | 7·0 | 14 |
| MARS | 142 | 687d | 24h 37m | 4,200 | 0·11 | 4·1 | 3·1 | 10/-70 |
| JUPITER | 485 | 11·9y | 9h 55m | 88,700 | 317 | 1·3 | 37 | -140 |
| SATURN | 890 | 29·5y | 10h 14m | 75,100 | 95 | 0·7 | 22 | -160 |
| URANUS | 1,790 | 84·0y | 10h 40m | 30,000 | 14 | 1·7 | 14 | -185 |
| NEPTUNE | 2,800 | 164·8y | 15h 50m | 28,000 | 17 | 2·2 | 16 | -210 |
| PLUTO | 3,700 | 248·4y | 153h | 3,600 | 0·8 | ? | ? | -220 |

*A table of planetary data*

reflected light with a spectroscope, and this has always been a difficult business, since corrections must be made for the absorption due to the gases in our own atmosphere. Recently a new spectrometric method has been devised, and this—Multiple Interferometric Fourier (MIF) spectrometry—together with space probes, is likely to yield much new information, and perhaps cause us to revise our ideas a good deal in the next few years.

The escape velocity is related to the mass. It is the speed that a rocket must attain (in the absence of air resistance) to escape from the planet's gravitational pull. A low escape velocity makes landing and take off easy, so that bodies with low values are the first choice as targets for space flights.

### The minor planets

Mercury is the smallest of the planets and closest to the sun. Although its orbital period is 88 earth days, it appears to rotate on its axis only once in 59 days. Its

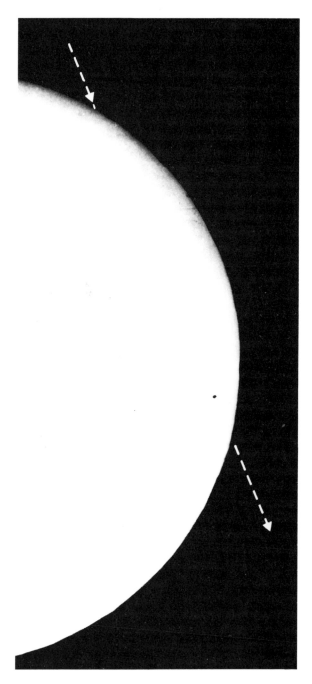

*Mercury, the innermost planet, shows up no bigger than a tiny black dot as it passes across the face of the sun*

*An artist's impression of the European Space Research Organization's observation probe scheduled to pass close to Mercury in 1975. In the background the path that the probe will photograph at close quarters is marked on a drawing of the planet*

sunlit surface is consequently extremely hot. In contrast, its dark side must be exceedingly cold, especially as there is apparently no atmosphere to trap warmth and convey it round into the darkness. It was once thought that Mercury had a trace of atmosphere, possibly even including vapours of metallic elements like mercury, arsenic and antimony, but recent studies have not detected any atmosphere at all. The planet's surface is distinguished by some vague, barely visible dark markings.

Mercury must, so far as we can tell, be a world without life. There can be no water there, for because of the virtual or total absence of atmospheric pressure it would vaporize and escape into space. Therefore no form of life known to us could exist there. Mercury is sometimes introduced into science fiction as a source of rare metals, but even if it were solid gold it would not be very attractive to astronauts, for the extremes of temperature would pose severe problems in exploring it. However, it would in theory be possible to visit its surface, because the escape velocity is small.

Venus is the mystery planet as far as the biologist is concerned, for it is shrouded in dense cloud, and we still know very little about its surface. Each new space probe may provide unsuspected surprises. Venus is only a little smaller than the earth, and of similar density, so that it is evidently of much the same composition. The day on Venus is about 120 of our days, though this comes about from the extraordinary retrograde rotation of Venus which, unlike any of the other planets, turns on its axis in the opposite direction to its orbital motion.

Venus has a dense atmosphere. It contains much carbon dioxide, and this has been confirmed by space probes. The rest is presumably nitrogen, with a good deal of water vapour. There have been reports, too, of traces of hydrochloric-acid gas. It is not yet clear what the clouds are composed of; the most likely possibility is that they are ice crystals floating high up in the cold stratosphere. There is a faint, unexplained luminosity of the dark hemisphere. Venus has little or no magnetic field. Occasional radio noise is heard from it, presumably from thunderstorms.

For many years the temperature at the surface of Venus was much disputed. It is now known to be very hot, because the dense atmosphere (and especially the abundant carbon dioxide and cloud) traps the sun's warmth in the same way as does the glass of a greenhouse. We know that the 'greenhouse effect' is pronounced, because the Russian space probe Venera 4, which impacted in October 1967, showed a temperature of about 400°C when its radio signals were lost, although it was probably still about 20 miles above the surface. At the same height, the pressure was around 10 atmospheres. There must be some difference between the temperatures during the long days and nights, but this may not be very great, because the dense atmosphere readily transports heat from the light to the dark side. The temperature must stay exceedingly

*The phases of the planet Venus as observed from the earth*

high all the night. The poles may be a little cooler, perhaps 200°C.

Clearly Venus is a very hot world. It is not yet clear if it is also a very dry one. One theory is that it is a desert, another (less likely) that it is covered with water, but in either case there are probably searing storms on its surface. If the temperatures and pressures are as given above, then liquid water might occur near the poles. Water boils at 312°C at 100 atmospheres pressure, but at the equator the temperature would be above the critical temperature for liquid water (374°C), so only steam could exist there. The surface is most probably composed of silicates of calcium and magnesium (these would be rocks like basalt) together with some silica (common sand) and mica. At high temperatures the carbon dioxide in the atmosphere would not readily form carbonates of calcium and magnesium. It is possible that hydrocarbons exist there as tars or asphalts. Radar reflections show that the surface is smoother than that of the moon. There is no evidence of oceans, but there may well be some high mountains.

The high surface temperature would not permit any life as we know it. There have been suggestions that

*The sensor capsule of the Russian space probe* Venera 4. *Immediately before radio contact was lost as it parachuted towards the Venusian surface in 1967, it recorded an external temperature of 400°C*

more moderate temperatures (perhaps a little below 100°C) might occur on the tops of high mountains near the poles, and it is true that the biological effect of high pressures is analogous to a moderate cooling. But even if we accept these, it seems very unlikely that any form of life could have arisen on Venus unless we consider exotic life-forms based on life chemistries different from our own.

More interesting is the possibility of life in the clouds of Venus. The upper layers of the atmosphere are evidently quite cold, and it is thought that the top of the cloud layer is well below freezing, about −40°C. A little deeper it is likely to be a little above freezing with a pressure of a few atmospheres. One could therefore imagine that some kind of life could exist floating in the upper atmosphere. The creatures would have to be small and light, and they would have to live in a narrow layer where the temperature was temperate and the pressure high enough for them to float on the wind. Either they would need a method of controlling their altitude as Sagan has suggested (perhaps by something analogous to the swim-bladder of fishes), or else they would have to multiply so fast that the losses due to some getting swept down into the scalding depths could always be replaced from the growing population. All in all this is not too probable, for indigenous life would have to have arisen under these unfavourable circumstances and within these narrow atmospheric limits. Yet there seems a real possibility that living organisms introduced by us might be able to colonize the upper atmosphere. If they were minute photosynthetic algae, they could convert the atmospheric carbon dioxide slowly to oxygen, with wide consequences. The planet would become more 'earthlike'. Less of the sun's heat would be trapped as the carbon dioxide became less, so it would become cooler, although still very warm to us. An ozone layer like our own would then form from the molecular oxygen. Although many terrestrial organisms can be carried through our own moist atmosphere, and very probably some continue to metabolize, we do not know if

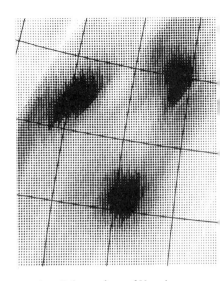

*A radio relief map of part of Venus's surface. The dark areas may represent mountainous terrain*

actual multiplication occurs in the air. But it seems quite likely that single-celled green algae can multiply slowly in damp clouds. If so, the clouds of Venus might suit them very well.

## The earth

Our own planet requires no extended description here. Yet it may serve to place the conditions on other planets in perspective if we briefly review the chief characteristics of our terrestrial environment. After all,

*The earth as it appeared to the first men on the moon homeward bound during the successful* Apollo-11 *mission*

life occurs at almost all the climatic extremes. If there is life on other planets we would expect it to be just as adaptable. The mean surface temperature on the earth varies considerably with latitude, though not so much with day and night. The average temperature at the poles is about $-30°C$, and at the equator about $30°C$. The seasonal changes vary from big swings of $50°C$ or more on the continents to only a few degrees on oceanic islands. The hottest deserts reach $55°C$. There is a definite 'greenhouse effect' produced by our atmosphere, because the mean surface temperature is $15°C$, instead of $-29°C$, as would be expected from its distance from the sun. The earth is probably unique in the solar system in having large bodies of liquid water as oceans, and is the only one so far known to have a strong magnetic field.

The atmosphere is about a quarter oxygen and three-quarters nitrogen with small amounts of carbon dioxide, water vapour, and rare gases like argon. Most of the oxygen has been produced by green plants, but it may be that some is accounted for by extensive loss of hydrogen in the past. The ozone layer blocks most of the sun's ultraviolet light. The earth's strong magnetic field produces the Van Allen belts of charged particles that girdle the planet.

We may note in passing that if we were on the planet Mars we should find it very difficult to determine the precise conditions that prevail on the earth with equipment such as has been available to us up to the present. Indeed, it has been said that Martian astronomer could only be sure of detecting life on earth by intercepting the intense radio transmissions due to our television programmes!

## The planet Mars

After the three inner planets, Mercury, Venus and the earth, there is a gap of 50-million miles to the next, Mars, which is therefore much colder than our own planet. Mars is smaller and lighter than the earth, and presumably for this reason it has lost most of its atmosphere. It is, however, rather puzzling that it

*Virtually the whole of the earth's atmosphere is confined to within ten or twenty miles of the surface. Here this comparatively thin atmospheric skin is caught and illuminated by the rays of the setting sun*

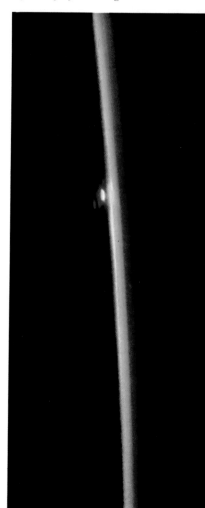

*Seasonal changes observed on Mars. During the Martian summer the north polar cap shrinks dramatically and the mysterious dark areas between the poles increase in prominence*

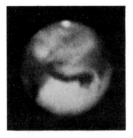

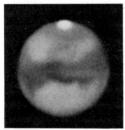

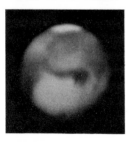

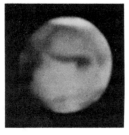

should have so little atmosphere, because it is massive enough to have held most of the heavier gases, particularly as it is not hot like Mercury. Mars, like the inner planets, is dense and composed of rock, and not, like Jupiter and the other giant planets, of gas and liquid. The escape velocity is fairly low, so that take-off from Mars will not require anything like as much energy as take-off from the earth. Mars has no magnetic field.

The Martian day is 24 hours 37 minutes, very close to our own. The surface is freely visible, and there are no oceans, for these would shimmer in the sunlight. Temperatures on Mars must be low for most of the time. The equator at noon can reach about 30°C, but at night the temperature falls to at least $-50$°C.

Recent studies by Mariner space probe indicate that the atmosphere is mainly carbon dioxide, with little or no oxygen (less than 0·1 per cent) and presumably some nitrogen and argon. Traces of ammonia and methane have also been detected. The pressure at the surface of the planet is only about one-hundredth of an atmosphere, and for this reason alone there cannot be any appreciable amount of liquid water on Mars, for it would swiftly evaporate. The atmosphere of Mars shows a 'blue haze', possibly due to very fine dust particles; but it has been suggested that this haze is produced by volatile products of some lowly form of plant life, just as blue hazes on the earth are due to terpenes produced by terrestrial plants. The nature of the Martian blue haze is still quite obscure. It seems to scatter blue light very strongly and to be in the lower atmosphere. However, it has recently been suggested that it is merely an effect caused by lack of contrast in surface features when viewed in blue light. The thin atmosphere would make an effective meteorite screen and absorb much of the solar X-rays. Clouds are sometimes seen. Those at high altitudes may possibly contain ice crystals; those at low altitudes are apparently dust storms. There have been reports of oxides of nitrogen in the atmosphere (and these are quite toxic to terrestrial organisms) but they are unconfirmed.

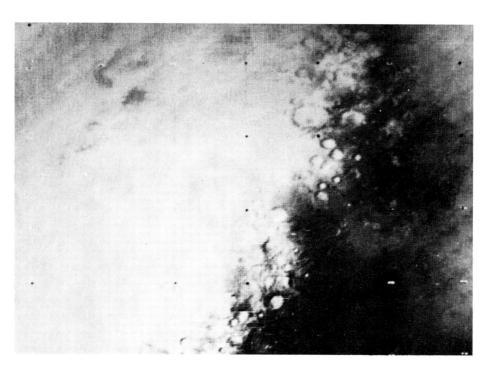

The surface of Mars is a reddish colour, and this is why it has the name of the Red Planet. There are two main sorts of hue. There are light reddish areas commonly called 'deserts', and darker ones called 'seas', though evidently both are deserts. The light areas match fairly well the spectrum of limonite, a brown, hydrated iron oxide. If limonite is predominant (no doubt together with silica sand) this would argue that an oxidizing, moist atmosphere has existed some time in the past.

In addition, Mars shows white polar caps, of which the northern one is permanent, though the southern often vanishes. These caps show seasonal changes, growing in the winter and shrinking in the summer. It is possible that they are composed of ice, or more precisely hoar frost, though it has been suggested that they are made of solid carbon dioxide. As the caps shrink in summer a dark band appears along the edge, just as though the frost had melted and dampened the

*Crater-scarred terrain is highlighted along the edge of the south polar cap in this photograph from Mariner 7. The cap itself is thought to be formed either of frozen carbon dioxide a few feet thick—or less likely—of ice barely an inch deep*

ground, and it seems that this would be possible. But the band may just be an illusion. The dark areas of the planet also show seasonal changes, which are one of the main pieces of evidence in favour of life on Mars, for this could indicate the growth and decay of some lowly plant life, perhaps rather like mosses and lichens. The dark areas become darker when the nearer polar cap is shrinking. That is, a wave of darkening occurs in the spring and summer in each hemisphere, and the wave passes from the polar cap toward the equator. This change in tone may be due to an increase in moisture in the surface dust, but no very plausible explanation on these lines is yet available. More persistent changes also occur. Since 1956, for example, a dark spot has been prominent in the Arcadian Desert.

The seasonal changes have been adduced as evidence for life on Mars. The thin atmosphere does not rule this out. There cannot be any large amount of liquid water on the planet, but it is quite possible that melted frost soaks into the powdery surface and forms a moist soil in which life would be possible. The latest data show the air to be at least ten times as dry as that in the Arizona desert. Yet there may be some deep depressions on the planet (perhaps the Trivium Charontis is one) where the climate would be warmer and damper than elsewhere. And if there is any volcanic activity this would provide local areas with warmth and moisture. Any plant life would of course have to be efficient at conserving water. Spectroscopic examination of the dark areas by Sinton showed absorption bands which he interpreted as being due to complex carbon compounds, which would of course strongly suggest that there were living organisms on Mars. This interpretation has been disputed, however, and at present no reliable evidence is available on the composition of the dark areas.

A point in favour of there being plant life on Mars is the behaviour of the particle size of the dust in the dark areas. This changes with the seasons in a way that is unlike a mineral. Also, after what appears to be a dust storm that covers a dark area with a layer of dust, it has

often been noticed that the dark tone quickly reappears, and this is very suggestive of vegetation. Polarographic studies suggest small micro-organisms, and these could be photosynthetic without necessarily liberating free oxygen (like some photosynthetic bacteria). We have noted, too, that there are bacteria that can exploit the small amount of energy in the change from ferrous to ferric compounds, and these compounds may be quite abundant on Mars. There are, however, some points against the vegetation theory. At mid-day the dark areas are hotter than the light ones, while vegetation would be cooler. Nor do the dark areas change more slowly in temperature during the day, as one would expect.

Ever since 1877, when the Italian astronomer Schiaparelli pointed out curious straight lines in the dark markings of Mars, these lines or 'canals' have figured in legend and science fiction. Percival Lowell of the observatory at Flagstaff, Arizona, believed they were irrigation canals built by intelligent beings, but he overemphasized their narrowness and straightness.

*The 'Red Planet'. Mars owes its alternative name to the reddish tint of its light-coloured 'deserts' composed, some believe, of silicates and iron oxides*

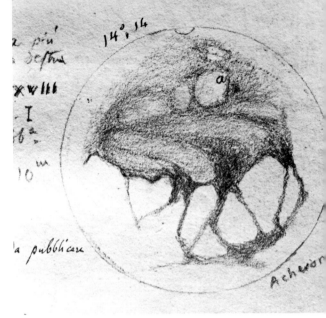

*In 1877 the Italian astronomer Giovanni Schiaparelli made a sketch (right) of the surface of Mars, drawing attention to what he called canali–narrow, dark-coloured areas which he took to be of geological origin. But in 1894 Percival Lowell, dubiously emphasizing the geometrical aspect of these features (below), concluded that they were built by intelligent beings*

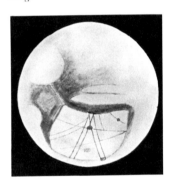

There certainly are some dark markings that are fairly straight, which join up the dark areas in a loose network. They, too, alter in colour with the seasons, and the astronomer Patrick Moore notes that they are very curious features, unlike anything else in the solar system. It is a puzzle that the Mariner photographs have shown no sign of them. They did, however, show many craters like those of the moon, which occur both in the light and dark areas. These photographs also show that the craters are very sharp. There cannot have been a dense atmosphere or running water for a very long time, otherwise the craters would show more evidence of erosion. There was little sign, either, of mountains or volcanic formations. These photographs have no great bearing on whether life exists or has existed on Mars.

If there have never been oceans on Mars, then perhaps life is unlikely to have arisen. On the other hand Mars might preserve remains of prebiological chemistry with clues to the way life originated on Earth. It may be that Mars had a dense atmosphere and liquid water some thousands of millions of years ago, and that the first steps in the emergence of life occurred before the atmosphere was lost.

The cratered appearance is somewhat against the occurrence of lowly 'vegetable' life on Mars, but by no means rules it out, for it might exist in some favoured spots. The lack of identifiable artefacts does not rule out intelligent life, for weather-satellite pictures of the earth seldom show anything recognizable as of human origin; an occasional field, road, canal or lumber camp has been seen, but very seldom. Nevertheless, the search for life on Mars is one of the most exciting programmes of present day science. For, as Philip Morrison of Cornell University has expressed it, it would 'transform the origin of life from a miracle to a statistic'. Traces of a Martian biochemistry would strongly suggest that life is likely to arise wherever favourable conditions prevail.

As we know something of the conditions on the surface of Mars, a number of investigators have simulated these, and looked to see whether terrestrial micro-organisms can grow under such conditions. These experiments have been referred to as 'Mars Jars'. Basically they consist of sealed jars containing sterilized soil, sand, limonite, etc. They are innoculated with organisms, and then the jars are filled with an atmosphere similar to that on Mars. The jars are alternately frozen and thawed to simulate the Martian day and night. Under such conditions several sorts of bacteria survive and some will multiply, though higher plants and animals perish. However, it is uncertain just how much moisture is present in the Martian soil, and this appears to be the critical factor. We have seen that some micro-organisms live in very saline waters, and that certain desert soils have high concentrations of deliquescent salts, with saturated salty solutions a few inches down in which microbes are growing. The Martian soil may be similar, particularly near the melting ice caps. There seems a real possibility that terrestrial micro-organisms could multiply in some parts of Mars. This question is more than an academic one, since we do not want to contaminate Mars or other heavenly bodies with terrestrial bacteria, if they could grow there.

*The recent close-range Mariner photographs – here covering more than a million square miles of the Martian surface – reveal no trace of Lowell's canals*

*Jupiter, the largest planet in the solar system, with a mass more than twice that of the other eight planets put together. Its surface is obscured by a turbulent, multi-coloured atmosphere thousands of miles thick. The famous Red Spot– first observed in 1878–is clearly visible, but its nature and causes remain unknown*

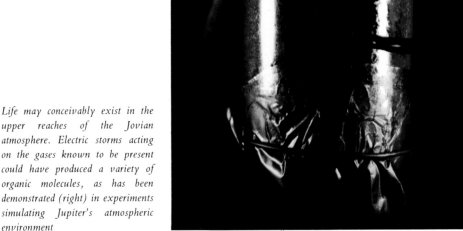

*Life may conceivably exist in the upper reaches of the Jovian atmosphere. Electric storms acting on the gases known to be present could have produced a variety of organic molecules, as has been demonstrated (right) in experiments simulating Jupiter's atmospheric environment*

## The outer planets

Jupiter is the largest of the planets, and is in some ways the most interesting. Its diameter is 88,000 miles, but it is of low density, only 1·34 times that of water. This at once indicates that it cannot be made up mostly of rock, and spectroscopy shows that its dense atmosphere contains much methane and ammonia together with hydrogen. This atmosphere is probably several thousand miles thick, and what we see is the cloudy turbulent lower atmosphere. Beneath this is probably a solid core, perhaps of metal or rock overlain by ice, but it is possible that it is composed throughout of gas, mainly hydrogen. If so, it would not be in the gaseous form, because of the enormous pressures; it would be more like a metal! The existence of semi-permanent markings suggests that there is a solid core, but the markings we see may be gas eddies produced by unseen surface features. The most prominent is the Great Red Spot. This behaves rather as if it were an island floating on the Jovian ocean, since it moves relative to other markings. It has been noticed, too, that the feature called the Southern Tropical Disturbance sometimes flows round the Great Red Spot as if the latter were solid.

So we know practically nothing about the surface of Jupiter. As one would expect from its great distance from the sun, Jupiter receives very little warmth. The temperature of the visible portions is very low, about −140°C, which is nearly as cold as liquid air. If the

*Saturn, circling the sun some 400-million miles beyond Jupiter, is a colder and even less hospitable world. Its unique ring system, comprising many billions of independently orbiting particles, measures 169,000 miles from edge to edge, but is probably no more than ten miles thick.*

deeper parts are as cold as this then much of the methane and ammonia must be liquid, perhaps solid. There would be an ocean of liquified gases with islands crusted with frozen gases, dimly lit, and swept by terrifying thunderstorms. However, it has been suggested that the cloudy atmosphere has a 'greenhouse effect', trapping both the sun's warmth and also heat from the interior of the planet (perhaps derived from radioactivity of rocks). If so, the surface might in fact be quite temperate. The atmosphere is too dense to allow determination by infra-red spectroscopy (as has been successful with Venus), so we shall have to wait until space probes can be sent there.

Both because of the ammonia, and also because the temperatures are probably so low, no form of terrestrial organism (or life constructed on the same plan as our own) could survive on Jupiter. But now that attention is turning toward the possibility of other forms of life with radically different chemistries, Jupiter is arousing a good deal of speculation, because it might support life based upon liquid ammonia, a possibility we shall consider in the next chapter. In any event, Jupiter should be exciting biologically, because it will probably show complex carbon compounds that have been formed by abiogenesis. There is plenty of energy available from ultraviolet light and lightning. There is already some evidence that hydrogen cyanide is present, which, as we have seen, readily forms more-complex organic compounds.

The enormous gravitational field would make Jupiter practically impossible to land upon, and so it is unlikely to be visited by man. It is not so much the low temperature and great surface gravity as the high escape velocity that will prevent this, because the amounts of fuel required by a rocket would be enormous if it were to escape from Jupiter's gravitational field. Jupiter emits radio waves from time to time, but the cause is obscure; intense thunderstorms have been suggested as an explanation.

Saturn is only a little smaller than Jupiter, and is best known for its wonderful rings, which consist of small particles (probably ice) rotating round the planet. Saturn has a composition similar to that of Jupiter; it has a deep atmosphere probably covering a core of rock and ice. The atmosphere, like Jupiter's, contains methane and ammonia, with rather more methane. The temperature is rather lower, and there cannot be much of a 'greenhouse effect', as it is so far from the sun. Some faint markings indicate atmospheric disturbances, but Saturn's surface is hidden by the dense clouds. It is another oxygen-less, frozen world, and its great mass and high escape velocity would, again, prevent us from visiting it. Any visits would be to one of its satellites.

Uranus, the third of the great planets, is very similar to Saturn, though smaller, and colder (about $-185\,°C$). Again it has a low density, and its atmosphere contains methane and a little ammonia.

Neptune is very similar to Uranus, with the same cold methane atmosphere.

Pluto, the outermost of the major planets, is so far away that very little can be seen. It is of course intensely

*Below left: Uranus—surrounded by an halation ring due to optical distortion—and four of its five satellites. Middle: Neptune and (arrowed) one of its two satellites. Right: photographs demonstrating the planetary motion of Pluto—first observed as recently as 1930*

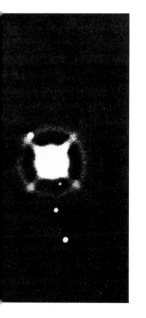

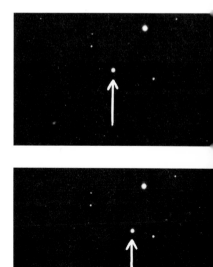

cold, but it presents several puzzles. It may be very dense, or perhaps much larger than it appears. We do not know anything about its composition or its atmosphere. Its orbit is very eccentric, sometimes coming within the orbit of Neptune.

## The moon

The moon, being so close to us, naturally attracts much attention, an attention greatly heightened by the recent triumphant exploits of astronauts. As far as life is concerned, however, it is unpromising to say the least. Waterless and without appreciable atmosphere, scorched on the sunward side and frozen on the other, it is a world where even the hardiest of micro-organisms would perish.

The moon is about 2,160 miles in diameter, only

*Man first set foot on an extra-terrestrial body at 0256 GMT on 21 July 1969 – the time and date of the first moon landing. Opposite: astronaut Neil Armstrong (who took the photograph) reflected in the visor of Apollo-11 crewman Edwin Aldrin. Right: A typical view of the inhospitable crater-scarred surface of the moon from an Apollo space craft in lunar orbit*

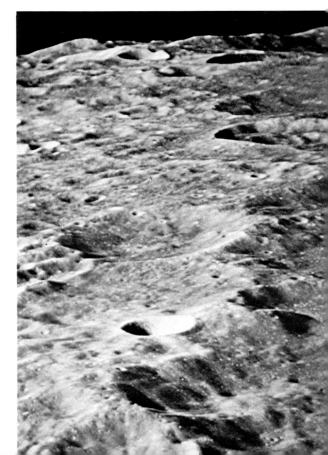

a quarter that of the earth, and is about 240,000 miles away from us. Tidal friction has stopped its own rotation so that it keeps one face turned towards us, and tidal friction is also slowing down the rotation of the earth at the rate of one five-hundredth of a second in every century. Over millions of years the slowing has been appreciable, and the daily and annual growth-rate patterns of fossil corals show that in the Devonian Period, some 380-million years ago, there were 400 days in the year. The day, therefore, lasted about 22 hours—in agreement with calculations from the tidal friction.

The results of the Surveyor landings suggested that the moon possessed rocks very like basalt, having a similar balance of elements to volcanic rocks on the earth, but the samples returned by the Apollo astronauts contain a higher proportion of some elements that are uncommon on earth and certain unfamiliar minerals. There is a layer of fine dust at least over the 'maria' (the flat plains which are miscalled 'seas'). Some of the dust is iron-containing and magnetic. There is still dispute over whether the craters of the

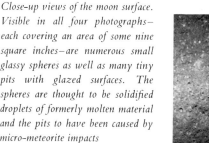

*Close-up views of the moon surface. Visible in all four photographs—each covering an area of some nine square inches—are numerous small glassy spheres as well as many tiny pits with glazed surfaces. The spheres are thought to be solidified droplets of formerly molten material and the pits to have been caused by micro-meteorite impacts*

moon are mainly of volcanic or meteoritic origin. Probably both processes have been active. Some evidence for recent volcanic activity has been seen.

Gravity on the moon's surface is only one-sixth of that on the earth, and the escape velocity is 1·4 miles per second. Owing to its nearness and low escape velocity, the moon provides an obvious first stepping stone for space exploration. In the 1970s it will become the site of man's first permanent scientific station on another world. This station will be equipped in particular for delicate astronomical observations which we find difficult to make through our own murky atmosphere, for the moon has no atmosphere to obscure the view.

*Earth's nearest neighbour the moon, almost certainly a lifeless world. Since its atmosphere is a near vacuum, its surface is exposed to constant bombardment by sterilizing solar radiation*

It will not, however, be an easy world to dwell on. The lack of atmosphere will make any surface station liable to puncture by stray meteorites. Stations would probably be dug in to avoid excessive ionizing radiation from solar 'flares' and the extremes of temperature (about 130°C in the sun and perhaps −180°C at night). A few yards down, however, the temperature is fairly constant and fairly cold. It has therefore been suggested that there might be ice some way beneath the surface, and, on the sunny side, moisture just above that. If so, micro-organisms might conceivably grow in this moist layer. Yet the likelihood of indigenous micro-organisms arising seems extremely improbable. It is, however, worth considering whether the moon could have trapped any living organisms (now dormant or dead) from space. These could have come from the earth, perhaps as the result of explosions caused by large meteorite impacts. They would be much less likely to have come from another planet like Mars, simply because of their dispersal through large volumes of space. But the moon may conceivably yield evidence in favour of the panspermia hypothesis, improbable though this is.

Of greater interest is the suggestion that the moon might have some organic matter dating from its early phase when it had an atmosphere and when solar

*An artist's impression of an envisaged permanent lunar base able to support a 12-man team and provide facilities for continuous research. Living quarters and laboratories are well dug in for protection against solar radiation, extremes of temperature and micrometeorites*

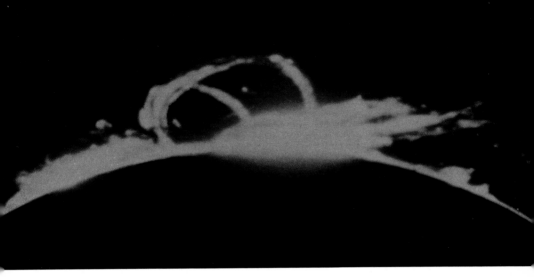

irradiation was synthesizing various organic molecules from methane, ammonia and water. If these molecules have been preserved they might offer important clues suggesting how life may have started on the earth itself. And it should be possible, too, to find meteoritic material that has undergone little change, and so help answer many questions about the origin of carbon compounds in meteorites, and whether meteorites have contributed much toward the formation of abiogenic organic compounds on our planet. So far, however, analysis of moon dust has shown very few traces of carbon compounds and no evidence of life forms.

*A solar prominence, or 'flare' – an arch of flaming hydrogen rising more than 50,000 miles above the sun's surface. Ordinary sunlight is masked off by an opaque disc*

## Satellites and asteroids

All the planets except Mercury, Venus and Pluto have satellites. Our own satellite, the moon, is the best known, but we have a fair amount of information about some of the others. None of them is at all likely to harbour any indigenous life, but as we may visit some of them in the future we are interested particularly in their temperatures, escape velocities and the composition of their atmospheres if they have any. The same applies to the asteroids.

The satellites of the planets are all small bodies,

and because of their weak gravitational fields all but one or two have lost any atmosphere they may once have had. Jupiter has four large satellites, the heaviest being Ganymede, with an escape velocity of 1·8 miles per second. Of Saturn's 10 satellites, one, Titan, is quite large, some 3,500 miles in diameter, and a thin atmosphere of methane has been detected, so there may be a trace of a methane-ammonia atmosphere on some of the other larger satellites, such as Triton, a satellite of Neptune. All the satellites of the great planets are very cold, but have low escape velocities, so they would make good staging posts.

The two satellites of Mars—Deimos and Phobos—are only a few miles in diameter, and the innermost, Phobos, rotates round Mars in less than a Martian day. It has also been reported to be slowing down at a rate that is much too fast for a dense body. This is still an open question, but it has been suggested by the Russian astronomer I. S. Shklovskii that Phobos is so light because it is hollow, and that it may be a space station of a long vanished civilization.

Most of the asteroids have orbits between Mars and Jupiter. Over 1,500 are known, and it is possible that they are the fragments from the disruption of a major planet that once existed between Mars and Jupiter. Of course, it may be that they are simply particles of the original material of the solar system that failed to coalesce. None of them is more than 500 miles in diameter, most being very much smaller than that. They therefore have no atmosphere, and are of little interest except possibly as space stations. Since they are so small, their rotation would tend to throw off any body resting on them near their equator. The theory that they were formed from the break-up of a planet has prompted the suggestion that some may be made of very dense metals like gold or platinum.

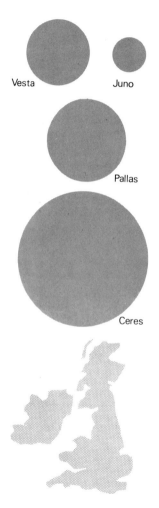

Vesta Juno

Pallas

Ceres

*The four largest asteroids and the British Isles drawn to scale. Asteroids are not necessarily spherical: many are known to be extremely irregular in shape*

## Organic matter in meteorites

Most meteorites are composed chiefly of iron, and a few are stony. Of the latter, a very small number contain large amounts of carbon compounds. As long ago as

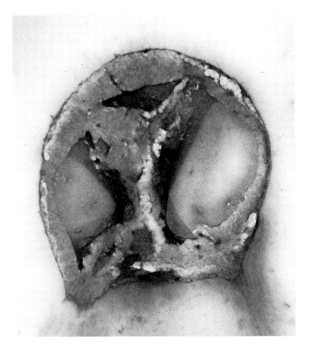

*This microscopic particle, found in the Orgueil meteorite that fell in France in 1864, is apparently 'organized' much as a cell is organized and forms part of the seven per cent of organic matter present in the original meteorite. But all attempts to establish it as evidence of extra-terrestrial life have so far proved inconclusive*

1834 the great chemist Berzelius tried to analyse one of these, but he could reach no definite conclusion on whether the carbonaceous matter was derived from a living source.

In 1961 B. Nagy and G. Claus of New York University reported that they had found organized bodies resembling micro-organisms in the meteorite that fell at Orgueil in France in 1864. A number of other reports have appeared, both on this meteorite and on other carbonaceous chondrites. Unfortunately, all these meteorites have long been exposed to contamination by terrestrial micro-organisms, and there is, moreover, considerable doubt as to the organized nature of some of these 'fossilized' microbes. There is some confusion, too, about the reliability of certain chemical staining tests that were believed to be specific for certain compounds of biological interest. Thus the Feulgen stain test, thought only to show up DNA, apparently can stain some minerals under certain conditions. Some of the organized objects appear to be contaminating pollen grains and the like. Bacteria have been cultivated from the Murray meteorite, but they were probably of terrestrial origin, as the meteorite is porous and lay in the earth for a while. Nobody yet seems to have tried to identify these bacteria by comparison with known soil bacteria.

Nevertheless these meteorites do contain complex carbon compounds, and some of them are very like the compounds in petroleum. It is believed that petroleum is derived from living organisms that were buried in ancient sediments, because it contains certain compounds, like phytane and porphyrin, which are chemical derivatives of the chlorophyll of green plants (including microscopic algae). But experiments on abiogenesis are now yielding so many compounds by purely chemical methods that it is becoming increasingly difficult to distinguish with any certainty between compounds derived from living and non-living sources. Recently J. Duchesne and his colleagues at the University of Liège have found strong evidence that carbonaceous chondrites contain indigenous organic molecules as free radicals which are unlikely to be due to heating or cosmic rays and may well, therefore, be of biological origin.

Two other puzzling pieces of evidence have been presented. Extracts of the Orgueil meteorite are optically active, and rotate polarized light to the left (laevorotatory), while samples from likely sources of contamination (dust, etc.) rotate it to the right (dextrorotatory). Since optical activity is only found in materials from living creatures, this points towards a biological origin for the complex carbon compounds in the meteorite. The opposite direction of rotation suggests a form of life unrelated to that on earth. But if the meteorite has been contaminated with soil bacteria and remained damp for any length of time, this could have led to very puzzling consequences. The bacteria might grow and produce compounds that rotate light to the right. But, as Sagan points out, they might also preferentially destroy the molecules that are dextrorotatory, and leave behind those that are laevorotatory.

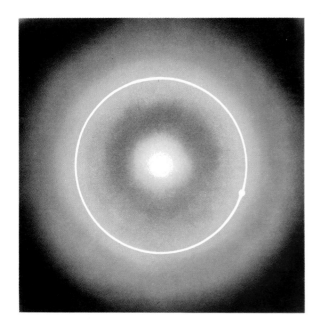

*Life as we know it requires a planet which, besides being massive enough to retain an atmosphere, must orbit its sun at a distance ensuring a tolerable range of surface temperatures. In our galaxy alone there are probably millions of planets orbiting in the 'life zones' of their respective solar systems*

If the meteorite compounds had been originally formed by a purely chemical agency they would consist of equal parts of dextrorotatory and laevorotatory molecules, but after bacterial attack the extracts might be either laevorotatory or dextrototatory.

The other piece of evidence is that the Orgueil meteorite contains porphyrin pigments, which are believed to be characteristic of materials of living origin. It should also be remembered that the Orgueil meteorite is rich in bound water, and it is somewhat of a puzzle where this could have come from, though water, in the form of ice, may exist on asteroids and in comets, and the meteorite may have been derived from one of these sources. If the carbon compounds are of biological origin where could they have come from? Some have suggested that the hypothetical planet that broke up to form the asteroids may have been a bearer of life, and the meteorites come from there. Others think that the moon may have broken away from the earth after life arose here, and that the meteorites are fragments derived from the breakaway of the moon. Despite enormous advances in analytical techniques in the last few years it is clear that conclusive evidence on the biological origin of indigenous carbon compounds in meteorites is still awaited. It is to be hoped that the search for evidence will be pursued, for as Urey has said, if proved, 'this would be the most interesting and indeed astounding fact of all scientific study in recent years'.

Other heavenly bodies as abodes of life require but brief mention. The sun, like other stars, is of course too hot for any material to remain either solid or liquid. When we look, therefore, outside the solar system we naturally think of planets belonging to other star systems. Nevertheless Shapley has suggested that there may be bodies in space that are warm enough to support life and not hot enough to be luminous. That all terrestrial life happens to depend on the sun is not proof that life could not use 'low grade' radiation energy. In fact some bacteria can live on chemical energy which might be supplied indefinitely by volcanic action. The search for very cool stars—stars that are barely red-hot—has begun, and there is evidence that they are not uncommon.

One further suggestion, by Freeman Dyson of Princeton University, is that intelligent beings might surround their dying sun with a vast spherical shell to trap all its energy, and live inside this globe. The captive sun hidden inside would have its energy converted to infra-red waves, which the sphere would radiate. For mechanical reasons, however, the shell could not be rigid; it would be more like a swarm of small space stations.

## Life-detection apparatus

A great deal of thought and planning is going into the construction of apparatus to detect life on the planet Mars in particular. Such apparatus poses a considerable challenge for its designers—it must be light and reliable, and it must be both sensitive and specific. The small payload available on space probes limits the weight, and extreme reliability is of course required on

The projected Voyager space programme will involve the landing of an unmanned space craft on Mars. Here a fully automated miniature laboratory—designed to take photographs, collect surface samples, carry out various scientific experiments, and transmit data back to Earth—unfolds itself on a desert hillside

*Television cameras prominently mounted on a* Surveyor-*type unmanned lunar landing craft.*

such expensive projects. The apparatus must be able to detect life in small amounts (if that is the word). It must also be capable of distinguishing clearly between the living and non-living.

The list of proposed detection systems is quite large and impressive, and includes the following (with the basis for the choice in parentheses): (1) television of the surface (characteristic forms of living things, like trees, plants, animals, etc.); (2) microscopy of surface constituents (if any life exists one would expect microorganisms at least); (3) gas chromatography of heated soil samples (complex organic molecules and fragments can be identified); (4) mass spectrometry of heated samples (as above); (5) ultraviolet spectrometry (the peptide bonds of protein have a characteristic absorption at 195 millimicrons wavelength, and other absorption bands of nucleic acids, etc., would be detectable); (6) J-band detection (the modification of the absorption spectrum of certain dyes when they react with large molecules of biological origin, like protein or DNA); (7) optical rotation (all living systems have a preference for either 'left' or 'right' handed molecules of sugars, nucleic acids or amino acids); (8) liberation of radio-isotopic carbon dioxide from culture media (micro-organisms should attack simple carbon compounds and liberate radioactive carbon dioxide); (9) measuring growth of micro-organisms by cloudiness in culture media (micro-organisms should make them turbid); (10) biochemical reactions for specific substances associated with living creatures (we may expect some of these to be present in other kinds of life. ATP is an example which can be detected in minute amounts by the luminescence it produces with the firefly luciferin system); (11) the catalysis of oxygen exchange between the heavy oxygen isotope (O-18) and water (apparently unique to living systems).

Many of the suggestions have been rather haphazard; workers have chosen convenient markers of life as we know it. The British chemist J. E. Lovelock has discussed more broadly the general strategies involved, arranging them into two main classes: the

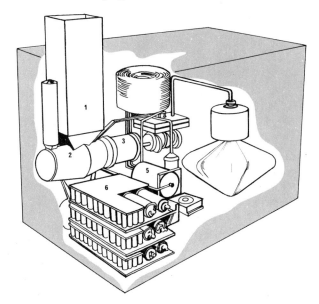

*A gas chromatograph capable of identifying organic compounds. A sample placed in the hopper (1) is vaporized in an oven (2), injected (3) through a long reactor tube (4) containing a special powder. As various gases escape they are detected (5) and recorded (6)*

*Below: a space microscope with no moving mechanical parts, designed to detect micro-organisms in soil samples. The sample particles are carried into the instrument by aerosol action*

search for order (which is equivalent to informational complexity), and the search for non-equilibrium (that is, self-maintaining systems which use energy to become relatively energy-rich, and are in metastable equilibrium). He suggests searching for large molecules of precise molecular weights (characteristic of living creatures) and heating soil in the planetary atmosphere to see if extra heat was given out (this would detect quite unfamiliar life forms, for it would indicate metastable equilibrium, and it could also be combined with gas chromatography).

Most of the life-detection experiments assume that any indigenous life on Mars will behave very like terrestrial organisms. This seems quite unsafe, and although there are sound reasons for looking for well-known systems (because we can evaluate any results with some confidence) it would be unwise to rely on them alone, or to assume that negative results rule out life of any kind. It is particularly difficult to know how to construct culture media for creatures of whose nutritional requirements we are quite ignorant.

The surface of Mars has already been scanned by television camera from a height of several thousand miles. No sign of life was detected, but it has been repeatedly pointed out that pictures of the earth's surface from such heights reveal little or no evidence of life on our own planet. A close-up view from a Mars lander would of course be more useful, although micro-organisms and lowly forms of vegetation would still escape detection. There are instruments planned to map the surface of Mars for traces of water, which would be particularly useful in the event of future landings.

The small vidicon microscope suggested by Joshua Lederberg would be much more promising. The main difficulties would be to scan sufficient dust, and some focussing device might be needed.

The gas chromatograph is quite a sophisticated instrument. The records of the traces would be transmitted back to Earth, and many chemical compounds could be identified with some confidence.

*The ingenious 'Gulliver' probe ejects long sticky threads attached to small projectiles. Dust particles and—if any are present—micro-organisms adhere to the threads, which are then reeled in and subjected to a chemical life-detection test*

Mass spectrometry would give rather similar information, and could deal with numerous samples.

The liberation of radioactive carbon dioxide has been the basis of the fascinating 'Gulliver' probe. This would throw out a sticky thread to catch dust, and reel it back through a chamber of culture medium, containing radiocarbon-labelled chemical compounds. Micro-organisms would release radioactive carbon dioxide which would be measured. The instrument is very sensitive indeed; tests on desert soils and bare rocks have shown clear evidence of the scanty micro-organisms present in these habitats. The chief difficulty is to know what labelled compounds would be attacked by the unknown life forms, and a wide range of likely ones may be chosen.

'When Professor Wolf Vishniac conceived a device to search for life in space it was inevitable that his biologist friends should name it the 'Wolf Trap'.' So reads a NASA report. The heart of the Wolf Trap is a growth chamber with sensors to determine cloudiness of the growth medium due to the multiplication of microbes, and a detector for the acid which would be produced by such growth. The media would be such that anaerobic growth would be possible, because Mars lacks free oxygen in its atmosphere. Dust would

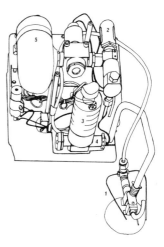

*The 'Wolf Trap'. A dust sample sucked through the collector (1) is forced, by gas passing through the regulator (2), into the culture chamber (3) where any micro-organisms introduced will grow and multiply. Consequent changes in the turbidity and acidity of the culture medium are recorded by the sensor unit (4). The large component (5) is the gas reservoir*

**123**

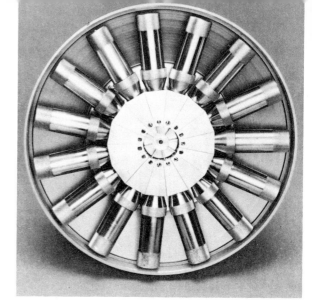

*The 'Multivator', which is able to subject equal portions of a sample to a dozen or more different biochemical tests. In particular it is designed to detect the enzyme phosphatase, which would almost certainly indicate the presence of life*

be collected by a jet of gas and carried to the growth chamber. Like the Gulliver probe, the composition of the medium will be quite critical, and it must be able to distinguish cloudiness due to growth from that due to the dust sample.

One of the instruments suggested for biochemical reactions is Lederberg's 'Multivator', in which a series of different biochemical analyses could be carried out. The enzyme phosphatase is widespread in terrestrial organisms and could be present in other life forms, so that detection by the Multivator of its action upon phosphate compounds would be a useful indicator.

## The probability of alien life

Until comparatively recently man considered himself unique, and had an anthropocentric and geocentric view of the universe. But far from being at the hub and centre of the cosmos, we in fact live on one small planet of a small and very undistinguished star. We now suspect, too, that the earth is not unique as an abode of life. The experiments which show that complex organic compounds (of the sort needed for primitive life to evolve) can be made artificially under conditions similar to early planetary atmospheres afford one piece of evidence. They imply that life may regularly arise under suitable conditions on other planets. The observations of colour changes on Mars is another point; it is possible that life has arisen independently there. Perhaps there are thousands or millions of planets with life on them. Because we know so little about the possibilities of life based on chemical systems other than our own, we will assume that we are considering in the main carbon-based life in a watery environment, with only a few comments on more bizarre life forms.

First we must ask what the chances are of life existing elsewhere in our own solar system. Apart from the possibility of ammonia-based life on Jupiter and Saturn, the only likely planet is Mars. Venus is too hot. But it is still very difficult to evaluate the Martian evidence, and conflicting views are held by knowledgeable investigators. It should not be many years, however, before we can test these experimentally, by means of Mars landers equipped with life detection apparatus, and, later, by human exploration. Until then we would be wise to suspend judgement.

Next we must consider the chances of life occurring in other solar systems in our own galaxy, the Milky Way. We shall, of course, be unable to make very accurate estimates; indeed one of the main points that will emerge is the difficulty of making estimates of this kind. It is nevertheless interesting to see what we would predict on the basis of our existing knowledge. The calculations involved are usually of the kind known as combinatorial. A simple example is to calculate the number of Scotsmen in Paris with beards and blue eyes. if we are told that there are two thousand Scotsmen in Paris, and one-tenth have beards, and one-quarter have blue eyes. The proportion who have *both* beards and blue eyes is expected to be about one quarter of a tenth, if there is no special predilection of blue-eyed Scotsmen to grow beards (or not to grow them). We would thus estimate the blue-eyed bearded Scotsmen in Paris to be about $2,000 \times 1/10 \times 1/4$, which is 50. In this sort of calculation we multiply the various individual estimates together.

Let us now apply this simple but reasonably reliable method to our problem, and begin by giving symbols to the various factors.

$N_L$—is the number of planets in our galaxy which possess life, which is what we wish to estimate.

$N_s$—is the number of stars in the galaxy which would be suitable suns for life-bearing planets (not all stars would be).

$f_p$—is the proportion of stars which have planetary systems.

$n_e$—is the average number of planets per solar system which would have a suitable environment for life.

$f_h$—is the fraction of these suitable planets where life has arisen.

The formula $N_s \times f_p \times n_e \times f_h$ therefore represents the number of planets possessing life, $N_L$.

We can now consider our estimates of each of these factors in turn.

$N_s$. Our galaxy contains something like 100,000-million stars, (i.e. $10^{11}$). Not all would be suitable for our present purpose. Many are too young, so that the planetary systems would be too hot, or too recently formed for life to emerge. Other stars are too cold. Yet others are double or triple star systems, the suns circling each other, and their planets would have erratic orbits, sometimes very close to the stars, sometimes distant. The planets would probably be alternately frozen and fried. Such stars would be unlikely to have life-bearing planets. Even so, there are many suitable stars which are single, hot enough, and old enough to have planets on which there has been sufficient time for life to evolve. For the earth, we believe that about 1,000-million years was the time required for life to arise. Estimates of $N_s$ vary a little, from about $10^9$ to $4 \times 10^{10}$. We will take it to be somewhere between these, say $4 \times 10^9$.

The value of the next symbol, $f$, is much more uncertain. Some astronomers believe that most stars have planets. The argument is that if stars condense out of clouds of interstellar dust, then they should be rotating much faster than most of them in fact are. Two-thirds of the stars are rotating far too slowly on this theory, and so is our sun. But in our own solar system the deficiency in the angular momentum is well accounted for by the rotation of the planets. This suggests that most other stars may have planets which account for the deficiency in rotation of the stars concerned.

Other evidence for planets is, however, very poor. There may be planets around two of the twenty or so stars that are nearest to us: Barnard's star (6 light-years away) and the star system 61 Cygni (11·1 light-years away), which have been detected by the slight wobble in the motion of the stars, or by dimming of their light as the planet moves in front of them. Some astronomers have estimated $f_p$ to be almost 1, that is 100 per cent, but we shall assume it to be about 0·1, since this is what two out of twenty would suggest.

As to $n_e$, we again have very little to go on. In our solar system there is the earth, and perhaps Mars, with a suitable environment. We might take one as a rough guide. However, two things are required here: a planet must be heavy enough to retain an atmosphere (otherwise its water would soon vanish) and it must also be within a narrow zone round its sun so that it is at a moderate temperature. Faint stars have narrow zones and there is only a small chance of a planet occupying it. Estimates are therefore lower than one, and we will assume that on average there will be only three suitable planets per ten solar systems. $n_e$ is therefore 0·3.

Our last item, $f_b$ is usually taken as 1. This means that every suitable planet will develop life within a reasonable time (say $10^9$ years). But we are here arguing from an example of one, our own planet. It would seem fairer and more realistic to admit that life may be an improbable event, though not extremely improbable. We might take 0·1 as an estimate.

Now, by substituting for the terms in our formula, we obtain the expression $4 \times 10^9 \times 0·1 \times 0·3 \times 0·1$. $N_L$

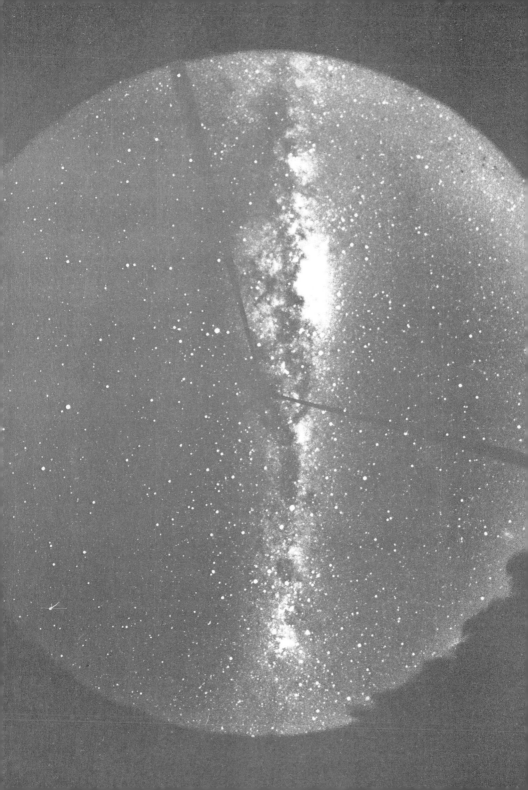

is therefore 12 million. We might accordingly expect that it is very likely that planets with other kinds of living creatures are really very common in the Milky Way. If we remember, too, that there exist *several million million* galaxies, it seems yet more certain that we are not alone in the cosmos.

However, as we shall see later, this does not mean that we could easily visit these other worlds. There are only about a thousand stars within a radius of 44 light years, and probably about a hundred are old single stars. If so, the chances of life within this distance would be rather less than 2 to 1 against. Clearly the chances of finding a planet to suit us as colonists without protective clothing would be a good deal less!

There is, too, an important difficulty with calculations of the sort described above. Each of our individual estimates is only an estimate, and should properly carry with it a figure representing our uncertainty about how accurate our estimate is. We can of course take our highest plausible figure for each, and then recalculate using our lowest plausible figures, but this is not a very suitable procedure, because one expects high and low guesses to cancel out to some extent. There are a good many statistical problems involved in doing this accurately, but, if we can make some simplifying assumptions, we can arrive at a very rough idea of the accuracy of our estimate that $N_L$ is 12 million.

Our calculations in fact suggest that there are probably not less than 200,000 and not more than 800-million life-bearing planets in our galaxy. The important point here is not how good our guesses are, but how *uncertain* the overall estimate becomes when we have much uncertainty in the individual estimates. This point will arise again later.

*Part of the Milky Way, which represents the great stellar disc of our galaxy viewed edge-on. How many of these stars have planets capable of supporting life? The author believes that a reasonably conservative estimate of the number of actual life-bearing planets in the whole galaxy would be of the order of 12 million*

# ALTERNATIVE BIOCHEMISTRIES 6

As we have seen, all living creatures on the earth are composed of complex carbon compounds, of which two kinds are always present: nucleic acids and proteins. The former provide the blueprint of inherited instructions, and the latter the tool kit for carrying out the vital processes. The solvent in which these vital processes take place is water. But the questions now arise: Need life always be based on carbon and water? Are nucleic acids and proteins the only compounds that can sustain living processes? We shall consider the second question first. Following the suggestion of N. W. Pirie, we shall call a form of life that is not of terrestrial origin an 'exobiont', and a form based on a different chemistry from ours a 'xenobiont'.

## One or many recipes?

There has been a good deal of argument over whether nucleic acids and proteins would be expected in exobionts. The reasons both for and against are not especially impressive. Some believe that the formation of life depends on well defined laws, and that life will soon arise in a suitable habitat. Others consider that life is rare in the universe, and therefore that it is chance rather than the nature of the vital compounds that determines its genesis. We have seen that simple amino acids and nucleic-acid derivatives are readily formed from methane and ammonia in experiments in abiogenesis. This may be taken as evidence that nucleic acids and proteins are the most probable vital

*All life on Earth is based on water and carbon—here represented by an ice crystal (opposite top) and a diamond*

compounds to arise, even if others are possible. But, equally, exobionts may more often employ *other* compounds from the abiogenic soup, and it could be *we* who are exceptional. We may note, too, that the twenty amino acids we employ seem to be an arbitrary selection of those possible. Effective proteins could presumably be made with a different selection. Again, it has been argued that nature experimented with many possible bases for life during the early history of our planet. Perhaps some of these did form incipient living systems, but once nucleic acids associated themselves with proteins they were so vastly successful that they over-ran, and finally gobbled up, what was left of the earlier eobionts. And if so, presumably the same may happen on other worlds.

This view, however, leads to a tangle of prickly arguments. If one set of polymers can supersede another, might not a new basis for life supersede us? Perhaps it could not do so on the earth, because there is no abiogenic soup left to start from (and any we may make in the laboratory is too small in quantity and has insufficient time, so we trust). But may not nucleic acids and proteins have been superseded on other worlds? If so, we must conclude that other systems could exist. And we might advisedly prefer to keep them at a distance. Perhaps it is more comforting to believe that nucleic acids and proteins are the unique lifebearers, that life arises only on planets very like our own, or even that we are alone in the universe. But all these arguments rest on very tenuous grounds. It could well be that in other primeval seas, with different constituents, other molecules could appear which possessed powers of replication or of catalysis. It could be that proteins and nucleic acids appeared on the earth because they were particularly effective in catalysing chemical reactions involving the most abundant energy-rich compounds that were then present (compounds like ATP). On other worlds, with other energy-rich compounds, the polymers that handled them most effectively might differ from those that we terrestrials employ.

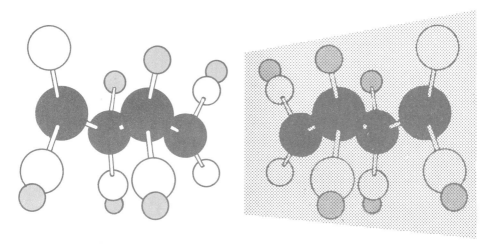

## Mirror creatures

Still, there should be variety in at least one respect. Terrestrial organisms use 'left-handed' or L-amino acids. Once a protein chain has been started with L-amino acids, the regular coiling of the chain into biochemically active molecules will be interfered with if a right-handed D-amino acid is introduced. One would thus expect that very early in evolutionary history the cell machinery would have been adapted specifically to using one particular form only; in our case the L form. But on another world the D form might become established, with right-handed protein helices. What would happen if D-protein organisms came into competition with us? Our digestive enzymes are probably inefficient at digesting D proteins, and their digestive enzymes might not attack our L proteins. Would this mean that we soon inter-permeated each other's societies, and even each other's bodies—each form immune to the other but also impotent to harm it? We would certainly compete for the optically inactive or neutral substances, like water, carbon-dioxide, light, and so on. Haldane, in one of his essays, suggested breeding bacteria to attack the mirror-image molecules of exobionts; we would hope that they would not do this themselves before we did!

We have supposed so far that water is the solvent

*Molecules can have either a 'left-handed' or 'right-handed' structure, the one the mirror image of the other. Terrestrial protein molecules are invariably left-handed, but proteins originating on another planet could all be right-handed*

of nucleic-acid protein systems. This may not necessarily be so. Certainly water is the most likely on chemical grounds, because the acidic compounds in proteins are hydroxyl compounds. A watery medium would limit life to the same sort of temperature range as ours, but, as we shall see, there are several solvents which might substitute for water at lower temperatures.

Some have assumed that nucleic acid and protein life would not only arise inevitably on suitable worlds, but that this life would evolve to give creatures very like those we know, not merely biochemically, but structurally; in other words, evolutionary laws are as consistent in their operation as physical ones. The trouble is that we do not know how much part is played in evolution by blind chance. We would, I think, expect there to be broadly similar classes of organism—photosynthesizers, herbivores, carnivores —but any similarity in detail would seem most improbable: we should not expect to find insects or whales. As Calvin has pointed out, there has not been time on our own planet for evolution to have tried all possible life forms. We know that the rate of evolution is subject to some serious limitations that make it a slow process even on a cosmic scale.

## Vital solvents

In 1913 a book by L. J. Henderson appeared called *The Fitness of the Environment*. It is now something of a curiosity, but whatever philosophical viewpoint it tried to present (this is by no means clear) it did discuss in detail the important and unique properties of water as the solvent for living systems. Water is a good solvent for a wide range of chemical substances, and it is an *ionizing* solvent—one in which the molecules of salts like sodium chloride dissociate into separate ions (in this case sodium and chlorine ions). Water is stable but not entirely unreactive. Another property now known to be of particular significance in biology, and responsible for a number of other properties of water (such as its high latent heat of vaporization) is that it is a highly associated liquid—that is, there are

*In combining to form a molecule of common salt a chlorine atom 'borrows' an electron from a sodium atom. The resulting sodium and chlorine ions, which are oppositely charged, attract each other to form the molecule, but dissociate in water*

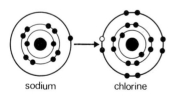

sodium                    chlorine

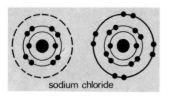

sodium chloride

loose bonds, called hydrogen bonds, holding the water molecules in loose chains and sheets. These hydrogen bonds are constantly shifting from one molecule to another.

These properties of water are of special significance because an ionizing solvent allows a wide range of chemical reactions to occur. The hydrogen bonding assists in holding biological molecules together while reactions occur. It is hydrogen bonding, too, which is responsible for the specific bonding between adenine and thymine, and between guanine and cytosine, which is the physico-chemical mechanism of the DNA helix and the genetic code. Hydrogen bonding also coats certain dissolved molecules with a sort of jacket of water molecules, and this, too, is sometimes important in biochemical reactions.

Although water is much the best known, there are a number of similar ionizing solvents to consider. Living systems might conceivably be based on any one of these ionizing solvents. But the great range of chemical reactions in cells is not restricted to ionic ones, while very weak ionic properties are shown by many other solvents not usually classed as ionized. G. C. Pimental suggests that hydrocarbon solvents could support life over a very wide range of temperatures, from the intense cold of liquid methane to hot tars and asphalts. An imaginative story on this theme described 'petroleum organisms' (Hal Clements's 'Critical Factor').

It would be advantageous for a solvent to have high latent heats of vaporization and crystallization, and a high specific heat, so that the effects of sudden changes of temperature would be minimized, and it should not be too viscous. We should also remember that under the high-pressure conditions prevalent on giant planets there may be no sharp distinction between the liquid and the gaseous state, at least if the temperature were above the critical temperature.

## Ammonia life

What are the likely alternative solvents to water? The most obvious is also the one which is present in the

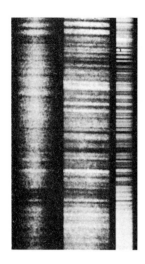

*The spectra of Saturn (left) and Jupiter, matched against that for ammonia gas (right), reveal the presence of ammonia in the atmospheres of both planets*

[1] ammonia $NH_3 \rightleftharpoons H^+ + (NH_2)^-$
water $H_2O \rightleftharpoons H^+ + (OH)^-$

[2] water
$H^+, (H_3O)^+$
$(OH)^-, O^{--}$

ammonia
$H^+, (NH_4)^+$
$(NH_2)^-, (NH)^{--}$
$N^{---}$

[3] $HCl + NaOH \rightarrow NaCl + H_2O$
$H.OH + NaNH_2 \rightarrow NaOH + NH_3$

atmosphere of some planets—liquid ammonia. The giant planets, Jupiter, Saturn, and probably Uranus and Neptune, have dense atmospheres of methane, hydrogen and ammonia. We cannot see the surface of these planets, but it is likely that oceans of these liquified gases lie beneath the clouds, with water in the form of ice, though some water will be dissolved in the oceans. On the colder planets, like Neptune, the ammonia may be solid.

J. B. S. Haldane speculated in 1954 on the possibility of life in liquid ammonia. The British astronomer V. A. Firsoff has since considered this in more detail, and has pointed out the close parallels between water and ammonia analogues. We must first rid ourselves of a misconception, that the words acid and base (or acid and alkali) apply only to the familiar acids and bases like sulphuric acid and caustic soda. Liquid ammonia is ordinarily thought of as a strong and caustic alkali. Yet if one dissolves ammonium chloride in liquid ammonia, the ammonium chloride behaves like an acid. So, too, does water. The words acid and base can therefore be extended to include a broader class of phenomena, and these can be illustrated in parallel by ammonia and water.

Molecules of an ionizing solvent dissociate into equal numbers of two different kinds of ion, acidic and basic. Only a very minute fraction of the molecules are split up thus at any one instant (about 1 in 20-million-million water molecules). These primary ions may combine loosely with undissociated molecules, and produce new ions. More rarely an ion itself may again split up, but for most purposes we may leave this aside. The ionization of water and ammonia is shown in the margin.[1] The hydrogen ion $H^+$ usually combines with another solvent molecule, and the $OH^-$ and $NH_2^-$ can dissociate again, so that we may have in these solvents the following ionic products[2] (those with a positive charge are 'cations', those with a negative charge are 'anions'). We can then compare two simple reactions[3] in which an acid and a base neutralize each other to form a salt and also a molecule of solvent.

In the second reaction we have water, H·OH (i.e. $H_2O$) acting as an *acid*, with caustic soda, NaOH, as a *salt*.

It so happens that many carbon compounds have 'ammonium analogues'. Some of these are shown on the right.

Indeed, it would be possible to build up the analogues of proteins and of nucleic acids (and other compounds). As far as our present knowledge goes we would expect them to be molecules potentially capable of forming the basis of a living creature.

One of the most curious things, pointed out by Firsoff, is that an important group of organic compounds, the ordinary peptides, could exist in an ammonia system without change. This is because acid amides (another group of compounds) can act in liquid ammonia as acids, so that in place of an amino acid like alanine—$CH_3CH(NH_2)COOH$—we could have alanine amide—$CH_3CH(NH_2)CONH_2$. Such amide molecules could condense to form polypeptide chains identical to ordinary ones (except for a final $CONH_2$ instead of $COOH$). Proteins are in fact often very soluble in liquid ammonia. This is of particular interest in connection with the evolution of life, because if life on the earth evolved originally in a reducing atmosphere of methane and ammonia, this would allow a smooth transition, at least for proteins, from an ammonia to a water system. In the giant planets, too, a little water in solution is likely to be present in the methane and ammonia seas, so that mixed ammonia and water reactions might occur. The extremely cold temperature of liquid ammonia is not itself a bar to life. Reactions would be slow or fast depending on the chemical bonds that were involved. The great pressures would not matter, neither would the great gravitational fields, since the xenobionts would float in the oceans.

It is instructive to view our own planet through the eyes of an imaginary scientist on Jupiter, and entertaining essays have been written on this subject. To Jovian astronomers the earth would be extremely hot, liquid ammonia could not be present, there would

| water analogue | ammonia analogue |
|---|---|
| methyl alcohol | methylamine |
| $CH_3OH$ | $CH_3NH_2$ |
| $CH_3COOH$ | $CH_3CONH_2$ |
| acetic acid | acetamide |

be almost no atmosphere and no life-giving methane. They might conclude that life could not exist here.

There are other solvents besides ammonia that merit attention, but not many are likely to occur in quantity on planets. Hydrogen fluoride (HF) is an excellent solvent in theory, and one in which organic compounds are stable, but it is unlikely to occur in the free form. Sulphur dioxide ($SO_2$) is a possibility. Hydrogen sulphide ($H_2S$) is the sulphur analogue of water, and a wide range of analogous reactions could occur. Analogues of proteins and amino acids are possible, and one amino acid, cysteine, has an -SH group, the analogue of the -OH group. Carbon chains would become increasingly stable at lower temperatures, and Firsoff suggests that nitrogen chains might become sufficiently stable to replace them. Hydrogen cyanide (HCN) is another promising solvent, and like hydrogen sulphide, traces may be present on Jupiter. Reactions in hydrogen cyanide might, however, be complicated by its tendency to polymerize.

## Why carbon?

The stupendous variety of compounds of the element carbon, and its willingness to combine with many other elements, provides the necessary basis for the complexity which is essential to living organisms. There is probably no strict dividing line between two important properties of carbon and their biological consequences, but we may conveniently think of them under two broad heads. Carbon can form long and complex molecules of great variety by virtue of the four available valency bonds of the carbon atom. This allows molecules of the necessary information content. But complexity in itself is not enough. Rubber, for instance, is a complex of an enormous variety of branched hydrocarbon molecules, but it is composed of carbon and hydrogen alone, and the chemical capabilities of hydrocarbons are limited. Life requires a variety of chemical potentialities. Some are involved in obtaining energy, others in storing or transporting

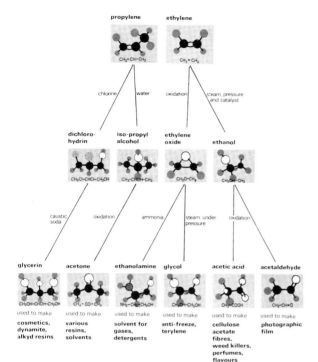

*The variety of possible carbon compounds is almost infinite. Here some useful derivatives of propylene and ethylene are considered*

it, others in synthesizing the building blocks of the cell (e.g. amino acids, nucleotides), and others in replicating the DNA and copying the genetic blueprint to give the working tools of the cell (proteins). This variety of chemical potential is easily obtainable, because carbon also combines readily and stably with many other elements. The most important besides hydrogen, are oxygen, nitrogen, phosphorus and sulphur, with a few trace elements (like iodine and iron) playing a sub-sidiary role. It is these elements that are active in the active centres of enzymes, and form, so to speak, the cutting edges of the tools. We know a little about some of these active centres, and can say, for example, that it is the iron atom held in a ring of nitrogens (supported, of course, by carbon atoms) that holds the oxygen molecule in the haemoglobin of our blood.

Now, very few other elements can even begin to compete with carbon in these two respects—ability to form large complex molecules and ability to combine with various other elements. For this reason there has been a widespread view that life could not be based on any elements other than carbon. This seems unduly pessimistic. At least twenty elements are handled by living organisms, and Pirie notes that early life forms may have been even more versatile, which suggests that xenobionts could be equally versatile, even to the extent of doing without carbon.

A valency of 4 is a great advantage in building large molecules, though this is possible also with valencies of 3 and 5. The elements that interest us most are those with a valency of 4 (like carbon, silicon, germanium and tin) or with 3 (e.g. boron, aluminium) or 5, (e.g. nitrogen, phosphorus, arsenic). A valency of 6 is also worth considering. (e.g. sulphur, selenium). The heavier elements, however, are less reactive and less versatile, so that it is the lighter elements like boron, nitrogen, silicon and phosphorus that are to be

*The ability of these red blood corpuscles to transport oxygen largely depends on the presence of iron and nitrogen in the haemoglobin molecule, but it is carbon that holds the molecule together*

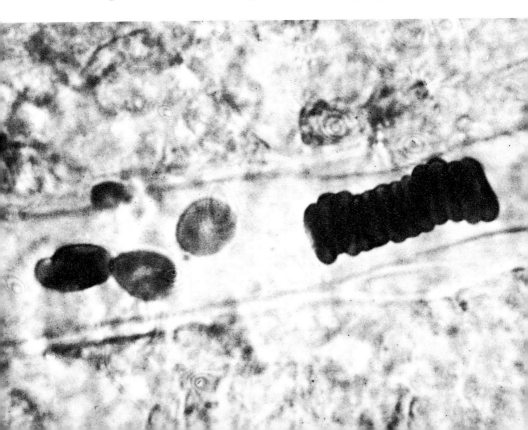

preferred. Carbon has both the best valency and the lowest atomic weight in its valency group.

## Silicon salamanders

Silicon (Si), like carbon, has a valency of 4, but unlike carbon it does not readily form long chains. This, however, is only true from our own special viewpoint, which assumes an abundance of oxygen. In such an environment silicon combines firmly with oxygen to give silica ($SiO_2$) which is what common sand is composed of. The silica also combines with other metals, like calcium and aluminium, to form silicates of various kinds. These silicates are indeed the major constituents of many rocks, but they tend to form large insoluble cross-linked molecules, which are not very suitable for building living organisms. If we consider other environments, however, the options are wider, particularly in the absence of free oxygen.

In 1909 a paper by J. E. Reynolds drew attention to the theoretical possibility that life might be based on silicon rather than carbon, and (because of the heat stability of silicon compounds) this would allow life at

*Silicon is similar to carbon in a number of ways, and is the second commonest terrestrial element. It combines readily with oxygen and occurs naturally as sand, mica and flints (above left). Above: crystals of pure silicon*

*Part of a silicone—a class of polymers with a backbone of alternate silicon and oxygen atoms*

$$\begin{array}{cccc}
CH_3 & CH_3 & CH_3 & CH_3 \\
| & | & | & | \\
-Si-O-Si-O-Si-O-Si-O- \\
| & | & | & | \\
CH_3 & CH_3 & CH_3 & CH_3
\end{array}$$

high temperatures. Haldane suggested that there could be life in the earth's interior based on partly molten silicates, obtaining its energy from the oxidation of iron. In recent years many new and complex silicon compounds have been made. The silicon polymers, known as silicones are already used extensively as lubricants and rubbers because of their heat resisting properties. These silicones have a backbone of alternating silicon and oxygen atoms, and each silicon carries two side chains, which are commonly simple carbon groupings like methyl ($-CH_3$). If the side chains were varied these molecules would be capable of carrying the large amounts of information needed for a living organism.

There are two special problems. The first is the heat stability of silicones. This is not very high, though higher than for carbon compounds. It depends principally on the silicon-oxygen bond. At about $350°C$ this bond begins to break and the fragments start to rearrange into rings, rather in the way that many carbon compounds (when heated without air) begin to form ring compounds like benzene. Therefore silicone-based organisms could not exist at temperatures above $350°C$. In fact the limiting temperature would be lower than this. It is not easy to say how much lower, for while it is a truism to say that a catalyst must be stable enough to catalyse its reaction before it is itself destroyed, this theoretical point is not easy to apply to living organisms which are a lot more complex than simple catalysts, and which must reproduce themselves before they 'fall to bits'. But by analogy with carbon we might estimate that 'silicone organisms' could 'live' up to $250°C$ and perhaps survive exposure to about $300°C$.

There are, however, other silicon compounds of greater stability. The silicon-carbon bond is stable to $500°C$, and certain aluminium silicone polymers are stable to $600°C$. There are also compounds with a backbone Si-N-Si-N. Many of these polymers are oils or rubbers, and are easily soluble in certain solvents, though they are often unstable in the presence of oxygen or water.

The other problem is that silicon does not form double bonds as carbon does. This means that its oxidation product, silica (nominally written as $SiO_2$), is really a huge lattice work of silicon and oxygen atoms. Instead of being a gas, silica is a rather insoluble solid. This is not an insuperable difficulty, for as A. C. Allison notes, the phosphate radical is not gaseous, and certain bacteria can metabolize insoluble substances like sulphur. The lack of double bonding, however, also limits the variety of compounds that silicon can form.

A few other classes of polymers have been studied by chemists. One of the earliest known was a 'rubber' of phosphorus, chlorine and nitrogen. Phosphorus-boron and nitrogen-boron polymers are very stable at high temperatures. Pirie has discussed germanium as a substitute for carbon, and there have been other suggestions, such as polymers of sulphur compounds, selenium, and so on. Firsoff notes that at low temperatures the noble gases like xenon form special co-ordination compounds. Solvents for high-temperature life have been little discussed, but molten sulphur, phosphorus sulphides and asphalt have been suggested.

In concluding this section on unusual 'life chemistries' we may briefly glance at another problem. This is the *rate* at which life processes occur. There would seem to be no particular reason to think this would be a limiting factor in life forms based on unusual chemistries. We might indeed at first find it difficult to recognize as living those processes that occurred slowly in very cold environments, such as liquid ammonia, or those that were rapid in hotter environments. But even at low temperatures some chemical reactions are rapid, and some reactions are slow at high temperatures. The significant point is that we would expect to find semi-stable chemical configurations with complex organization, whatever the rates of reactions. The main objection to very slow and sluggish life is that evolution might be so slow that there would be insufficient time for it to make much progress before the environment was destroyed.

Man is extremely proud of his intelligence. He is *Homo sapiens*, Man the wise. He is proud, too, of his humanity, his imagination, his creativity, his spirituality. It is not easy, therefore, to step outside ourselves and view these qualities in a detached way. But if we are to consider the possibility that intelligent life really does exist elsewhere in the universe and is not simply a subject for science fiction, we must make an effort to do just this.

## Bug-eyed monsters

Science fiction writers have imagined a range of intelligent creatures, some of them such travesties of gorillas, spiders and centipedes that they are affectionately known to the craft as Bug Eyed Monsters— B.E.M.s for short. Others are so man-like as to be called humanoids. In much of this writing it is clear that the stories are simply westerns in a space setting, and that the scientific content is negligible. But there have been some extremely imaginative attempts to consider the physiology and psychology of extraterrestrial sentient beings, such as the 'hnau' of C. S. Lewis.

One of the early pioneers was H. G. Wells, who as a biologist was aware that biological explanations are as necessary in science fiction as physical explanations. In the *War of the Worlds* he described the superspecialization of the brains of his Martians (compared

*Opposite: superintelligence in a fictional monster of another world is represented – spectacularly, if unoriginally – by grossly enlarged cerebral hemispheres*

with the rest of their bodies), and returns to this theme in *The First Men in the Moon*. Olaf Stapledon did the same most convincingly, in *Last and First Men*, with his 'Fourth Men, The Great Brains'. One of the few entirely original ideas supported by some plausible physiology is that of Fred Hoyle in his novel *The Black Cloud*. Hoyle imagines a vast interstellar cloud, composed of dust, ice, and various chemical compounds. The analogue of the brain is a region containing neurological structures laid down on solid bodies and having the ability to receive and transmit radio waves. Gas streams carry 'food' substances through the cloud, which obtains a store of energy from a star by surrounding it and absorbing its radiant energy. Since it can replace any part of itself as it wears out, it is potentially immortal; it may also reproduce itself by implanting the seeds of a similar organization in other suitable clouds. Only the nature of these seeds is left without some explanation.

But to return to organisms of a more familiar nature, exobionts will no doubt possess various bodily organs, just as we do. Their bodies will have to perform similar functions to our own, and these functions will be most efficiently carried out by specialized tissues. We would expect them to have digestive and excretory systems, a transport system or circulation to move nutrients around the body, and specialized organs to facilitate movements, biosynthesis and reproduction. Division of labour among the various parts of the body is very striking in living organisms—even in the simplest forms of life, once we look for it. Additional organs are required to control and co-ordinate the whole. With us this is partly achieved by a system of chemical messengers (hormones), and in plants such a system seems to suffice. In animals the main control is exercised by the nervous system (which can respond with more speed and versatility than circulating hormones) and this very likely arose because of the need to control locomotion as it evolved in the primitive animal stock. Free movement requires close control. Again in response to movement the more sophisticated sense organs evolved. It was no longer sufficient to be simply aware of wet and dry, of light and dark, of stillness and shock, of the tasty and nauseous. Very probably the emergence of the habit of predation forced the pace. Once there were creatures that both moved and devoured other moving creatures it became increasingly important for them to find their prey, or to avoid their enemies, more actively than plants, which simply grow up again after being grazed down.

It is quite possible that not all inhabited worlds have developed along lines resembling ours (one could easily imagine a world with only vegetable life), yet if any intelligent creatures are to evolve one would expect some rather similar series of evolutionary pressures to operate. It is difficult to see how a vegetable world could develop intelligence, for what needs would vegetables have that could cause the emergence of a brain? So a world with 'hnau' would probably have not simply the two classes of creature necessary to maintain a living world, that is, photosynthesizers (or others that use non-living sources of energy) and those that rot down dead organic matter. Their world would also have other familiar ecological classes, like harbivores, carnivores, parasites and symbionts. As to the morphology of exobionts, mechanical laws would operate on them as on us. Creatures on massive worlds with strong gravitational fields would have to be thick-set or flattened (at least if they were not marine), while flying creatures would have to be small, even if the atmosphere were dense. Conversely, on small planets with a low gravitational field, exobionts could be thin, spidery or flying creatures. Life in oceans would be less affected, of course, as individuals would be buoyed up by the fluid.

It is quite probable that their sense organs would be remarkably like our own. There are, after all, only a limited number of ways in which light or sound can be efficiently detected. There are only a few ways to focus light, and resonance is the best way of distinguishing sounds of different pitch. Moreover the size of a sense organ must be related to the length of the waves it detects. A visual image as sharp as those formed by our eyes cannot be produced in an eye one-hundredth of the diameter of our own, because the highest resolution is at very best no smaller than a quarter of a wavelength of light. Since the only radiation that freely penetrates planetary atmospheres is visible light and radio waves, exobionts would no doubt develop eyes at least, and these would be comparable in size with our own. In theory they might 'see' radio waves, but they would need enormous 'radio eyes' to form images of radio waves.

## Unfamiliar modes of perception

Though exobionts would probably be able to see, there are many other ways of perceiving one's surroundings, even for highly active creatures. They might have a very different view of the world from our own. Human life is strongly oriented toward the visual.

*A blindfold porpoise finds its way between underwater obstacles by emitting sound pulses and interpreting the echoes it picks up*

*Bats, by employing a rather more sophisticated form of sonar than the porpoise, are able to construct finely detailed 'sound pictures' of their surroundings. Left: a multiple-exposure photograph of a bat catching a mealworm. Below: the fish* Gymnarchus niloticus, *which locates its prey by detecting disturbances in an electric field which it generates itself*

Other animals have other dominant senses. As is well known, bats rely almost entirely on sonar; they hear the reflections of sound waves emitted by themselves as high-pitched squeaks. Since they can easily avoid even tenuous obstacles like telegraph wires they must be able to form something like a 'sound picture' in their minds. Some cave-dwelling birds (oil birds) have very similar powers, and there is evidence that whales and porpoises also have a form of sonar.

Yet more extraordinary are certain fishes allied to the electric eels. Instead of producing an electric shock to stun its prey, the fish *Gymnarchus* uses weak electric discharges to sense the whereabouts of other fish. It lives in a dark muddy world quite unlike our own, and yet, like bats, it avoids obstacles with ease.

*The rattlesnake is equipped to 'see'
its prey in the dark. Between eye
and mouth on each side of its head
is a pit lined with cells sensitive
to infra-red radiation—which is
emitted by all warm-blooded animals*

There is also a little evidence that some animals are sensitive to magnetism or radio waves, and there seems no theoretical reason why this should not be so. Rattlesnakes have special infra-red detectors in the pits on their snout, and this allows them to detect the direction of warm blooded animals. Perhaps they have a heat picture of the world as well as a light picture.

## The evolution of intelligence

As the elaborate control required for motion assumed greater prominence in higher animals, the nervous mechanism became concentrated in a brain. The development of a brain allowed, in due course, intelligence to develop in the higher vertebrates. Intelligence is a matter of handling very complex information, and this is what the brain is uniquely fitted to do. Sir Julian Huxley believes that the evolution of intelligence and consciousness is a major step, comparable to the development of many-celled animals from unicellular micro-organisms. He therefore views man as the first of a new kingdom of living things, the Psychozoa, while the more intelligent vertebrates are transitional forms. It is fascinating to speculate on what the later members of the Psychozoa may be like. Is it even possible that some of them may be machines?

This raises the question of just what intelligence is. The dictionary definitions are, variously, 'the faculty of reason', 'the capacity to adapt to new situations', 'the power of understanding', 'the power to acquire and use knowledge', 'the ability to invent, create and imagine'. By these criteria many animals besides man show a degree of intelligence. Even a plant can count after a fashion! The Venus Fly Trap has trigger hairs on its trap-like leaves which close on unwary flies if they touch the hairs. But the first touch does not operate the trap: the second does. The Venus Fly Trap counts quite accurately up to two. It would perhaps be better,

*The hinged leaves of the Venus Fly Trap capture a fly. The fact that they can only respond to a second stimulus, and not to a first, enables them to distinguish to some extent between living prey and inanimate objects*

therefore, to think of intelligence as a varied collection of different aptitudes, involving planning (forethought), memory (afterthought), learning (adaptability), and creativity (imagination, image-forming or pattern-forming). Perhaps, too, we should demand some minimum evidence that all of these abilities are present before we call it intelligence; or rather more precisely, say that the level of intelligence is primarily determined by the ability that is the least well developed. It is perhaps characteristic of intelligence that it operates by using a sequence of explicit tests of success: 'when I have achieved *this*, then I will do *that*'. The tests are explicit enough, for they lead to decisions, even when the subject is handled in an intuitive way. Thus creative artists, although they may not be able to explain the criteria of artistic success, are seldom in doubt that they are testing their creations against some standards of this kind, and constantly do this as they follow their inspiration. Whether this use of success criteria is restricted entirely to intelligent beings may be questionable, but it is certainly a prominent feature of intelligent behaviour.

Intelligence requires a large and elaborate brain. This is not entirely a matter of size. The brain of an elephant is larger than that of a man, but it does not have the complex convolutions which seem to be associated with greater intelligence. The elephant is one of the most intelligent mammals, but its brain shows great specialization of a low type.

Whales, however, present a striking contrast. Their brains are both large and convoluted. As far as structure goes they appear to be the equal of our own. Indeed, whales and dolphins (the latter are just small whales) are exceedingly intelligent, and are currently the subject of a great deal of research. There are few animals that can handle two different objects simultaneously, as do the dolphins at the Miami Seaquarium. Whether the whales and dolphins are inherently intelligent, and simply require teaching, or whether they lack some vital part of the complex needed for high intelligence (such as the ability to

*Dolphins—now known to be among the most intelligent of mammals —have unexpectedly advanced brains. Their remarkable capacity to learn and perform complicated tasks and tricks renders them of great interest both as objects of scientific study and as entertainers*

handle a complex language) is a matter of some interest. Man himself is not very intelligent before he is taught—indeed our experience of the few cases of children who have been brought up by wolves and later rescued, shows that they rarely attain anything like a normal intelligence, but remain at a mental age of a few years, little superior in intellect to an ape. For all we know whales may be extremely intelligent in some ways, and stupid in others. They may, for example have great abilities in underwater choreography, with whale ballet rivalling Covent Garden. It is also uncertain what drives whales have: we all know ourselves that keenness is the greatest asset in any mental task. What arouses keenness in whales? Our greatest difficulty is to imagine what their world is like, and what they might be thinking. This is difficult enough with familiar pets like dogs and cats, and despite the passionate protestations of animal lovers,

**153**

*Improvement in manual dexterity – the co-ordination of brain, eye and hand – may have been a key factor in the evolution of human intelligence*

we have very little knowledge of what goes through their minds. The imaginative effort is great even when, as in Stapledon's *Sirius*, we cast a human mind into the frame of a lower animal.

Intelligent creatures need brains, then, but is this all they need? Many anthropologists believe that human intelligence only evolved because of the interplay between brain, eyes and hands. The binocular vision of monkeys encouraged manual dexterity, and early men were cleverer with their hands than apes; this challenged their minds and encouraged the survival of the more dexterous and intelligent. For these reasons, it is said, intelligence of a high order cannot be expected from creatures without hands (or similar appendages). Thus the dolphins can never become really intelligent, while elephants, which have a trunk, might do so. All this does not prevent degeneration of limbs and sense organs *after* the evolution of a highly intelligent brain, though it would be difficult to think of circumstances which favoured this. However, dolphins and whales have a reputation of being at least as intelligent as monkeys, despite their lack of hands. If manual dexterity is a prerequisite of intelligence, one would not expect dolphins to be clever at all. It is of course true that dolphins do not have a technology, but it would be hard to argue convincingly that all higher intelligence in the universe must necessarily be technological. Rather more plausible is the argument that intelligent beings must be social beings, because only a society will produce the necessary teaching and the conflicts of interest that goad minds into thought.

Intelligent creatures could be large, especially if they were marine. But could they be very small? Could they be Lilliputians, or like those in John Wyndham's story 'Meteor', only the size of ladybirds? There seems to be no theoretical reason why nerve cells could not be much smaller than ours. The ultimate limit would be set by the limit of packing information into matter, for example by displacing individual atoms in a crystal according to some pattern. The DNA helix does not approach this limit, but even it is extraordinarily compact. The information content of the pollen grain of a lily is as extensive as a library's.

We have already considered the chances of life arising in a suitable environment elsewhere in the universe. Now we may ask whether, where life exists, intelligence always evolves if there is enough time? If it does, we would expect the incidence of intelligent life on other worlds to be quite high. On the other hand, it may require an unusual combination of circumstances to encourage the evolution of intelligence of a high order, and we have little evidence to help us answer this. Life had been evolving for at least 3,000-million years before the human race appeared; birds and mammals took a little less (only about 200-million years). Yet in the million years or so of human evolution our species has advanced in intelligence to a far greater extent than that to which mammals advanced in relation to reptiles over two-hundred-million years. We have virtually no clues as to why this was so, or as to precisely what combination of events caused such swift intellectual evolution. Is it that such favourable circumstances are very rare, or that we just happened to be the first mammals to develop intellectually? We suspect that there may well be groups of animals or plants which have a built-in bar to intelligent development. It would be hard to see how grasses could ever evolve brains. No creature seems to have evolved a true rotating wheel as an organ of locomotion, or caterpillar treads. Perhaps, then, other life systems are unable to evolve intelligent forms however much time they are given (and this is not unlimited, for planetary life-times are probably unlikely to exceed 10,000-million years).

We have also to ask ourselves what selective advantage intelligence confers. It is not simply in numbers or in mass: mammals are outweighed by grasses and men outnumbered by rats. It does confer on the species the power to live in a greater range of habitats, and consequently to survive greater dangers of natural catastrophe. What animal besides man lives both on Baffin Island and in the Kalahari? It is not this, though, for which we value our intelligence, but rather for the greater freedom of choice it gives to both individual and race, to please ourselves and to give full play to our curiosity and versatility—a risky luxury that our duller cousins cannot afford. We prize the powers it confers: to make great paintings, poems and symphonies, to fly to the moon, to wonder at the night sky.

## Could machines live?

So far we have been considering life in its obvious and traditional sense. But what about robots? Could machines be in any sense alive? Could machines evolve and develop into creatures that were living in all the ordinary senses of the word? Notice that the emphasis here lies upon the word 'could'. No machines yet invented, or imagined in any but the most sketchy and anthropomorphic terms (or better, perhaps, imagined in the likeness of human minds, rather than of human bodies), can claim to be either alive or intelligent except in the most trivial senses. Yet if there are advanced civilizations elsewhere in the universe, their machines may be vastly superior to our own. We should therefore explore what might come about given unlimited knowledge and skill.

The British scientist I. J. Good has proposed an argument based on the notion of an ultra-intelligent machine. This is a computer that can perform every intellectual function better than any man. As he puts it: 'Whatever we can do, it can do better. There are no ultra-intelligent machines on the earth at the moment; or, at least, if there are they have not told me of their existence. In my opinion, the first ultra-intelligent machine will be built on earth during the next fifty years'. The point Good is making is that if one can build a machine as intelligent as a man, it would be possible to set it to work to construct an even more intelligent machine, and furthermore this second machine will not cost much more than the first. The second machine can construct a yet more intelligent one at only moderate cost, and so on. This is 'Good's Law'. The argument is logical and tightly argued, but there are three difficulties.

*These 'solid logic modules'—miniaturized components of a modern computer—only mimic brain cells: the most advanced computer yet built does not compare in complexity with even the most primitive animal brain*

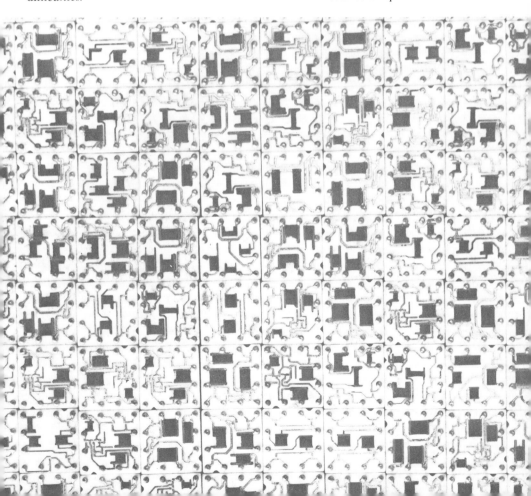

First, the cost may rise at a prohibitive rate, so that economic factors are limiting. This could certainly be serious if the demand for mechanical components should increase too steeply. Anything that increases as a factorial, for example, soon becomes prohibitive: if intelligence was measured on some hypothetical but realistic scale and shown to increase as the inverse factorial of the number of nerve cells in the brain, this would set severe limits on the intelligence of organisms. Our own brains, with about $10^{12}$ nerve cells would score around 15 on the intelligence scale, since factorial 15 (i.e. $1 \times 2 \times 3 \ldots \times 15$, written as '15!') is about $10^{12}$. A mouse, with a brain weighing a gram or two, and about $10^{9}$ nerve cells, would score around 12 on the intelligence scale (12! is almost $10^{9}$). But a superbeing whose intelligence was as great compared with us as our intelligence is compared with that of a mouse, would have to score 18, and this would need 18! nerve cells, which is nearly $10^{16}$, and its brain would weigh ten tons! Even taking intelligence at its simplest, as problem-solving, we note that the number of inter-relationships that must be explored to reach a solution increases steeply with the number of factors in the problem.

The same consideration would probably apply to intelligence in robots, and, if so, it would make even rudimentary intelligence extremely difficult to confer on machines. We should bear in mind, however, that we do not know very much about the relation between intelligence and number of nerve cells or electrical components, and it is at least possible that the increased technological power from ultra-intelligent machines would achieve an economic breakthrough by allowing manufacture of smaller and cheaper electrical circuits.

The second problem, and perhaps the most serious, is that ultra-intelligence may require a degree of logical power which we cannot incorporate into the first 'intelligent' machines. Like a critical temperature required to set a match on fire, there may be a minimum 'intelligence' without which the process of development cannot start. We are hampered here by our lack

*This computer, programmed to play draughts, is capable of improving its game, learning by its mistakes*

of knowledge of how our minds work, in particular the imaginative and creative sides of our nature. It is possible to think of machines solving algebraic theorems (and indeed some interesting experiments are being conducted on these lines), of playing chess (perhaps brilliantly), and so on. It is difficult to conceive of them composing great music or great poetry, or questioning the nature of the universe. At present most (if not all) of the 'intelligence' of computers is simply that of the human programmer embodied in his program. Yet I do not see how we can rule out the *possibility* of creative intelligence in machines, even if we do not see how to achieve it.

The third problem is rather like the second, but embodies a different bar to success. The rate of improvement of intelligence in each succeeding machine may be so slow that the process will be ineffective from the practical point of view. This might appear to be only a theoretical difficulty, were it not for the fact that we know evolution in living organisms to be a very slow process. Some of the limitations of

*The attempt by robots to supplant man is a favourite theme in science fiction. Above: 'cybermen' – on location near St Paul's Cathedral, London*

organic evolution will not apply to machines, but the basic difficulty will remain. This is that any change in an evolving system (that is, in a succession of descendants) must be relatively small and very carefully 'chosen', if it is not to make the system less efficient than it was. Tinkering with organisms is like tinkering with a watch. We must ensure that the modifications are at least sometimes successful, and success may be extremely difficult to define. This certainly applies if our tests of success are to be too difficult for human beings to comprehend. In some way, too, the machines must always be protected against obsessive circular arguments of the kind described in G. R. Dickson's story 'The Monkey Wrench', in which a computer is put out of action by the paradox: 'You must reject the statement I am now making to you, because all the statements I make are incorrect'.

## Consequences of Good's Law

Nevertheless, let us accept Good's argument and see where it leads. A civilization that invented ultra-intelligent machines would presumably arrange for these machines to be capable of making copies of themselves. Whether we allowed them to do this is another matter, because the machines might then 'take over'. But in principle the machines would be able to reproduce. They would contain a blueprint of their own structure, including a copy of their program—that is, their 'intelligence'. They would consume energy ('feed'), have some envelope separating them from the environment, and be in metastable equilibrium. Of course, they would, by definition, be intelligent, creative, sentient beings. They might grow and repair themselves. In what way should they not be living? It seems that we would find it extremely hard to find any absolute distinction between them and living organisms.

The interesting thing about putting the development of intelligence first is that ultra-intelligent machines could then evolve 'backwards'. They could use their intelligence to improve and reconstruct their own physiology. Living creatures first evolved powers of reproduction and feeding, then mobility and complex organs, and lastly intelligence. But ultra-intelligent machines could develop in the reverse order, once they had obtained their intelligence initially from us. They could first develop improved sense organs. Then they could become mobile. Lastly they could acquire independent power supplies of their own and re-produce themselves at will. We might be tempted to view this as retrograde development. We might aim

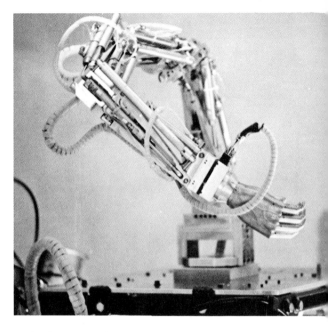

*A robot hand and arm which, in conjunction with a television-camera 'eye', is able to grasp and manipulate various objects on instructions from a large computer*

at higher mental or spiritual powers. They, on the other hand, might consider that the earthy was what was needed to complete their natures, that passions and parts would make them truly mechanical. And we, often afraid of the cold inhumanity of machines, could not with consistency simply condemn their efforts to become more like us.

They might of course supplant their creators. Indeed, one of the earliest of the robot stories, and that which gave us the word 'robot', was Karel Capek's 'R.U.R.' (Rossum's Universal Robots). In this story the robots do destroy mankind, but they also discover love, and so become 'human'. Capek's story, and other early tales of automata, have interesting historical connections with the 'Golem' of Jewish tradition, automata constructed by learned rabbis. If robots were to supplant us they would need to have a desire to do so. Perhaps we should heed a warning — in an article by W. T. Williams in *Nature* in 1963 — which is pertinent to the theme of evolution of machines:

Computers, being only brains, live in a cold world, free from all such desires and free, therefore, from the desire to think; and we shall be safe from their domination so long as we continue to build them solely as brains. If we go further—if we aspire to give them reproductive systems, digestive systems and all the facilities for sensuous enjoyment, then they will no longer be computers but genuine robots; and the world of Karel Capek, Isaac Asimov and Arthur Clark will be around us at last.

Isaac Asimov's celebrated Laws of Robotics, however, represent an attempt to implant moral values

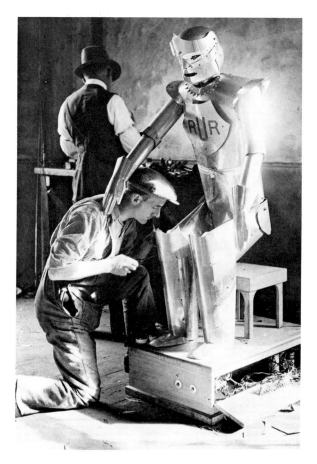

*A film-set version of one of 'Rossum's Universal Robots' which, in Karel Capek's famous story, dis-place mankind, only to become 'human' themselves*

in machines so that they would not supplant us:

(1) A robot may not injure a human being, or through inaction, allow a human being to come to harm.

(2) A robot must obey the orders given it by human beings except where such orders would conflict with the First Law.

(3) A robot must protect its own existence as long as such protection does not conflict with the First or Second Law.

It may be that among the galaxies there are civilizations where machines have supplanted their makers. Perhaps, too, the machines themselves died out because of some fatal flaw in their natures. It is, of course, evident that if we admit the possibility of mechanical life we do not need any longer to restrict possible environments to ones that resemble our own. As long as it was not so hot that solids melted, machines could operate and adapt themselves to these conditions.

Robots also face us with some difficult ethical dilemmas. Should we allow them to supersede us? They would not be 'flesh and blood'—not bone of our bone and flesh of our flesh—but they would be mind of our mind, and in this sense they would be our children. If man should love his creations, his creatures, his children, should we not love them, even if they supplant us?

## Human biomachines

But is this antithesis between living and mechanical a real one? Would it not be possible, and also advantageous, for composite entities—'biomachines'—to develop? Their range of capabilities would be greater, their capacity for enjoyment deeper, their desires better controlled, their lives longer, perhaps potentially immortal and infinitely perfectable. The fear of the human mimic, the cold machine, the machine-in-man's-clothing, may be misplaced. Perhaps it is in symbiosis that our joint future lies. Symbiotic organisms can be extremely successful. As we have seen, the lichens are all symbionts. They consist of a protective fungus and a photosynthetic alga, and they are successful in the harshest of habitats: their reproductive parts are constructed to carry both partners in their joint enterprise. We have also suggested that we ourselves and all other higher organisms may perhaps be symbionts of eucaryotes and bacteria. Perhaps machines *can* become human. Indeed, Good suggests that biomachines may be more sympathetic toward other forms of life than wholly biological organisms would be. Their very intelligence might make them so, as is well expressed in the following quotation from Stapledon. 'The Fourth Men, the Great Brains' (who were in effect just thinking machines) began to realize the limitation of their insight into values. Stapledon writes:

> Should they concentrate their efforts upon the production of new individuals more harmonious than themselves? Such a work, it might be supposed, would have seemed unattractive to them. But no. They argued thus. 'It is our nature to care most for knowing. Full knowledge is to be attained only by minds both more penetrating and more broadly based than ours . . . To refrain from this work would be irrational'.

*In this scene from the film 2001 an astronaut is seated before the 'eye' of an ultra-intelligent computer which secretly plots his destruction – on the ground that the non-rational element in human behaviour is likely to jeopardize the success of the mission*

# SPACE TRAVEL

On 12 April 1961, the Soviet cosmonaut Major Yuri Gagarin became the first human being to complete one circuit of the earth in an artificial satellite. On 23 December 1968, the American astronauts Colonel Frank Borman, Major William Anders and Captain James Lovell left the earth's gravitational field, entered that of the moon, and achieved the first lunar circumnavigation. On 21 July 1969, Command Pilot Neil Armstrong stepped on to the moon's surface. Man has taken his first decisive step away from his planetary home; no longer is he restrained by the earth's gravitational field. The pioneer spacemen are the forerunners of a multitude. What may their successors accomplish?

Space travel can be divided into two broad areas. The first has to do with travel within our solar system. We already possess much of the technology for this, and journeys to the planets are largely a matter of time and money. Travel outside the solar system, however, involves great problems which we have not yet solved, and some (such as travel to other galaxies) may prove to be insoluble.

## Exploring the solar system

The basic machinery for space flight is by now familiar to most people. Propulsion is given by a rocket, which carries the payload, a space capsule. The rocket may consist of several stages, all but the first riding on the nose of a larger one. The giant three-stage Saturn-V

*In July 1969 man was on his way back to the earth (opposite) after walking on the moon itself, yet only eight years had passed since Yuri Gagarin (above) became the first man to orbit the earth*

*The gigantic Saturn V rocket slowly easing its 3,000-ton bulk upwards a few seconds after the Apollo-11 lift-off. Had there been an emergency during launching, the small rocket attached to the nose cone of the command module would have lifted the astronauts clear of the Saturn's booster stages*

*Below: the relative depths of the gravity 'pits' of the moon, Mars, the earth, and Jupiter (abridged). The figures indicate the velocities in miles per second that a rocket must attain to 'climb out of' the pits*

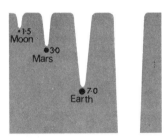

rocket is 364 feet high and weighs 3,100 tons. The first stage delivers a thrust of 3,750 tons. Yet the capsule that carried the first astronauts round the moon weighed a mere 6 tons. This vividly illustrates the major problem of space travel—that of weight. The usual chemical fuels are capable of only a limited performance: the payload can be no more than a few per cent of the starting weight. The bulk of the starting weight consists of fuel that is used up—by the Saturn first stage, at a rate of 15 tons per second—in driving the payload into orbit and then away from the gravitational pull of the earth. The force of gravity dominates space travel. It is as if objects upon the planets were at the bottom of deep pits. The greatest expenditure of energy is in climbing out of these 'pits'. Once out, progress is relatively easy.

We cannot expect any marked future improvements in the power of chemical fuels. Those already in use are

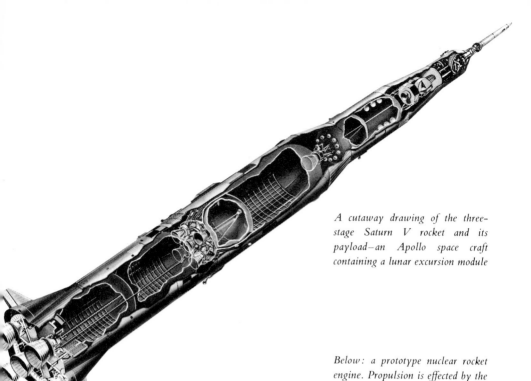

*A cutaway drawing of the three-stage Saturn V rocket and its payload—an Apollo space craft containing a lunar excursion module*

*Below: a prototype nuclear rocket engine. Propulsion is effected by the rapid heating of hydrogen as it passes through the reactor core*

near the useful upper limit for storing energy in chemical form. Only a small gain could be obtained by using air-breathing engines such as the ram jet, which obtains much of its fuel (the oxygen, in fact) from the earth's atmosphere. Nor is the development of nuclear-powered rockets likely to lead to substantially heavier payloads. And anti-gravity devices seem to be impossible on theoretical grounds. Forms of energy other than chemical cannot produce the thrust needed to boost a rocket from an earth lift-off to the required escape velocity of 7 miles per second (25,000 miles per hour). So we must assume that space travel in the foreseeable future will remain very costly and payloads small. Considerable savings in rocket fuel would be possible, however, if the fuel could be manufactured from local raw materials on the moon or the asteroids, which have comparatively low escape velocities.

*An artist's impression of a manned Mars vehicle designed to transfer a crew (in the two cylindrical cabins) from Earth orbit to Mars orbit in 140 days. The craft is propelled by rocket motors which act by accelerating hydrogen ions in an electric field*

Once a spacecraft is in orbit round the earth, however, there are other forms of propulsion available. These are ones in which the thrust delivered is small but the speed of the expelled propellant very high. This makes economical use of the limited weight of propellant that can be carried. If, for example, one accelerates individual atoms of an element in an electric field, they can reach very high velocities. Though the thrust is very small, one uses up very little of the propellant every second. Such a device could eventually drive a rocket at speeds far higher than would chemical fuels. But the small thrust (only 20 pounds per engine has so far been achieved) would mean that acceleration would be low: the electric rocket motor would only be useful for long journeys.

Since space is airless and very cold, elaborate space vehicles and space suits are required. Much advantage

*Above: technicians peering intently into a vacuum chamber observe the performance of an ion-accelerating electric rocket motor*

*Left: a nuclear rocket engine undergoing trials at a test bed in the Nevada Desert*

would be gained by having a manned orbiting station as the take-off point for long journeys. These would probably rotate slowly, and be constructed in the form of a wheel, to provide artificial gravity. Unlike excessive gravitation (a man can tolerate about 15 g lying down), weightlessness, or zero g, appears to have no serious ill effects on man, at least over short periods. Weightlessness is none the less a considerable nuisance: objects float about unless held down, liquids break into a fine mist and tend to cling to surfaces and spread over them, and drinking incautiously could make one choke. Also, after some days, weightlessness produces effects similar to those of being confined to bed, namely muscular weakness and, in time, increased fragility of the bones. These effects are due to lack of

*The cramped interior of the command module of an Apollo space craft as it appears during the final stages of the countdown and during the ascent into orbit*

*A manned orbiting space station designed to rotate and so provide crew inside the outer rim with a substitute for gravity in otherwise weightless conditions*

exercise in an environment where only slight exertion is needed. Astronauts avoid this particular problem by exercising with spring apparatus.

Space travellers also face other hazards. There is a small risk of collision with a meteorite, but on present evidence this is less than one in ten thousand for a week's voyage. More serious are the effects of solar flares. These flares are gigantic eruptions from the sun, often associated with sunspots. They shoot out streams of fast-moving atomic particles which produce ionizing radiation. This can be severe, and is also difficult to shield against. The earth's atmosphere normally protects us from these particles, but the shielding needed for an astronaut would be unacceptably cumbersome. Luckily the flares only last a few hours with long periods between them. The Van Allen radiation belts around the earth are soon crossed, and do not seem to be a serious hazard. On long

*A demonstration of the effect on liquids of a loss of pressure: at a simulated altitude of ten miles a glass of cold water rapidly boils away. Were the pressure-suit not worn, the subject's blood would similarly have boiled*

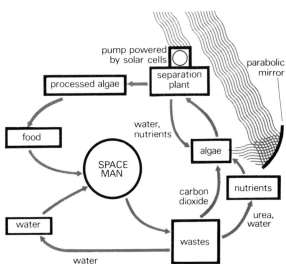

*A diagram showing how waste products can be recycled to provide fresh food and water*

voyages loneliness and boredom would become a serious problem.

Because of the problem of small payloads, much thought and experiment has gone into self-regenerating systems for food, water and oxygen. On a trip to Mars, for example, which would take over a year, it would just be feasible to carry all the necessary food, water and oxygen, but it would be much better if water could be recovered from the breath and excretion products, and if the carbon dioxide exhaled could be reconverted to oxygen. This can be done. Human volunteers have spent many months in a closed system of this kind. The energy required could be obtained either from the sun's rays, or from a small nuclear reactor carried on a long arm to prevent excessive radiation from reaching the astronauts.

It would, of course, be even more useful if the food could be recycled. The usual proposal is to illuminate cultures of minute green algae with sunlight. The algae grow and absorb the carbon dioxide, and are then used for food. Unfortunately the algae tried so far are very unpalatable, and it is said that animals will starve rather than eat them; and some algae produce traces of poisonous carbon monoxide. The algal cultures may also become infected with unwanted microbes. It is clear that we are still some way from a

*Volunteers in a sealed chamber fitted with closed life-support systems, observed during a 60-day period of isolation*

practical system. Moreover at distances beyond Mars, sunlight becomes too faint to stimulate good plant growth (though light-collecting mirrors might be used to overcome this).

## Journeying to the stars

Travel outside the solar system poses quite different problems. The time for a journey to even the nearest star would be prohibitive unless velocities could be attained approaching the speed of light itself. This again raises the problem of providing an adequate fuel supply, even in the case of the most efficient kinds of propulsion we can imagine, such as the conversion of matter directly into light energy. No material object can travel faster than light, and at speeds close to that of light a disproportionate amount of energy must be supplied to achieve a very small increase in speed. This is a consequence of Einstein's deduction that mass becomes infinite at the speed of light. Consider a nuclear fusion rocket of the future which, by converting hydrogen into helium, might accelerate a space ship to a speed 99 per cent that of light, on a flight, say, to a star 10 light-years away. The amount of fuel would have to be about a thousand-million times the weight of the astronaut, so the rocket would have to be as big as a hundred-*million*-ton ship. Even with the theoretical upper limit (a fuel of equal parts of matter and antimatter) the rocket would weigh more than the largest oil tanker afloat today. Turning round and returning to the earth would need more energy, because slowing down from these tremendous speeds would itself require an enormous fuel supply. It is, however, conceivable that a space ship could scoop up hydrogen from interstellar gas clouds as it travelled along, and use this hydrogen as fuel. If matter could be turned directly into energy an asteroid might be captured and propelled along, being gradually consumed as fuel. Dyson has suggested that a space ship could gain a tremendous boost from the energy of rotation of a binary, or twin-star, system, by making a loop round one of the pair. Especially interesting is the way the

*Left: a drawing which suggests the configuration of a theoretical 'photon' space craft. Photons emitted during a controlled reaction in which matter is annihilated by anti-matter particles would be beamed backwards by a gigantic parabolic reflector perhaps as much as a mile in diameter. Such a craft would be potentially capable of speeds approaching that of light*

acceleration would be unnoticeable to the astronauts involved because they would remain in a state of weightlessness throughout. This may seem surprising, but since the gravitational acceleration would act equally on all the matter in the space ship, they would not experience any acceleration relative to the ship itself.

## The aging of astronauts

The daily life of the astronauts in such space ships would perhaps not be much more difficult than on long interplanetary flights or long sea journeys. There have been suggestions that one day space travellers might be transported in a state of suspended animation, perhaps frozen in some way, or that feeding might be dispensed with by some method of generating ATP electrically within our bodies. However, such techniques may not be necessary. Einstein's theory of relativity predicts that an astronaut will age less quickly than those left behind. On a voyage to a star 10 light-years away it might take about 25 years to make the one-way trip; the extra time would be spent in reaching almost the

*During future space flights lasting several years or longer it may be necessary to conserve life resources by inducing a state of hibernation in personnel not needed on the outward or return journeys. In this fictional scene (right) the mummy-like apparatus near the top of the picture maintains and regulates the greatly slowed-down metabolic processes of an enclosed crew member*

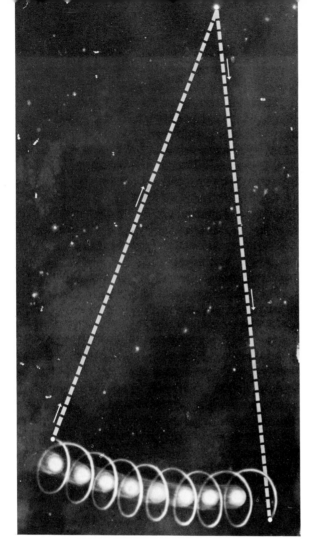

*Relativity theory suggests that as a space vehicle approaches the speed of light, time will 'slow down', for its crew without their being directly aware of it. Astronauts returning home after a journey to a nearby star (left) may find that they have aged considerably less than their erstwhile contemporaries*

speed of light and in slowing down again. But the astronauts would only record about eleven years as passing! So the astronaut's wife would welcome back a husband who had scarcely aged at all!

This effect, known as time dilatation, could in theory be exploited a good deal. Given enough energy to accelerate to the necessary speeds, a space ship could travel great distances within our galaxy in the lifetime of one of the travellers. E. Sänger calculates that by maintaining an acceleration of 1 g the ship could even reach other galaxies in 40 years by the spaceship's clocks. The trouble would come when the astronauts returned, for the earth would be millions of years older. Conversely, the astronaut would be like the Flying Dutchman, practically immortal by the earth's standards.

Perhaps we can envisage whole colonies of human beings journeying through space (and we should not be too cynical, for as Buckminster Fuller reminds us the earth itself is a spaceship), even if we cannot yet see how interstellar travel will be achieved. Indeed, it has been suggested that the entire earth might be propelled through space to find a new sun when our own sun begins to die. It would, of course, mean that we should have to live in roofed-in parts of the earth, artificially warmed and lighted. It may even be that speeds greater than that of light are not impossible, but that we do not know how to achieve them. Time-travel and space-warps are favourite themes of the science fiction writer; all we can say is that we do not see at present how they could be possible. Travel to another galaxy would demand something of the sort, however, for the galaxies are millions of times farther away than the nearby stars that we have been considering.

## Problems of contamination

Space travel brings with it a new problem, that of contamination of one world by another. There is now a good deal of awareness of this, and the rocket-launching nations now co-operate through appropriate international bodies (such as COSPAR, the Committee for Space Research) to keep this under review. This interest, slow to be aroused, dates from a paper by Lederberg and Cowie in *Science* in 1958. It argued that we know nothing of life forms beyond the earth (our biological knowledge is quite unlike our knowledge of physics in this respect); that theoretical biology is hard to develop without such knowledge; and that it would be a scientific catastrophe were we to deprive ourselves of the chance to study life on other planets by carelessly contaminating them with terrestrial organisms. This warning arises from the fear that terrestrial bacteria might exterminate any exobionts they come into contact with. But, even more important, if we do not know what our micro-organisms could do to other worlds, neither do we know what theirs might do to us.

Contamination may be considered under two headings: chemical and biological. Chemical contamination is not without significance, but it is biological contamination that is our main concern. This is because of the power of living organisms to multiply. A single microbe could in theory infect an entire planet, for if it grew at all, even if its growth rate were very slow, it would not be long before it was widespread. Chemical contamination is nothing like so serious a problem, because ordinary chemicals do not multiply. There should be no difficulty in avoiding the accidental release of toxic or radioactive materials on other planets. Some concern has been felt that the exhaust from rockets might interfere with both chemical and biological experiments, but in general the amount of contaminating material would be too small to be a severe nuisance.

There are, too, further reasons for guarding against the contamination of other planets. Apart from not wishing to destroy any indigenous organisms, we must be sure that any organisms we may find are indeed indigenous, and are not contaminants introduced by us, perhaps on earlier space probes. Then again, if there are complex organic molecules on other planets, these might represent the pre-biotic stages of the evolution of life. We would not want introduced micro-organisms to destroy these compounds before we had time to study them and perhaps form a clearer picture of how life arose on the earth. Furthermore, the introduction of microbes might produce pronounced changes in the chemistry of a planet's surface and its atmosphere. If we should wish to conduct experiments of this kind they should at least be planned, controlled experiments.

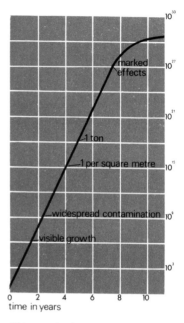

This graph shows the increasing rate at which a lifeless planet might be colonized by micro-organisms, assuming original contamination by a single cell and division into two daughter cells at an average rate of only once a month. Within two years, more than a million organisms would exist

At present most concern is felt about Mars. It is possible that some form of life exists there, and, as we have seen, some bacteria can multiply under conditions similar to those on Mars. The situation with Venus is less easy to evaluate. The surface is apparently too hot for terrestrial organisms. There remains, however, the possibility that single-celled green or blue-green algae might colonize the upper atmosphere and change it into an oxygenated atmosphere like our own. The other heavenly bodies concern us less. The moon is unlikely to be habitable by any terrestrial organism, while the other bodies are either much too hot or too cold. It is reassuring that both Americans and Russians are taking considerable trouble over sterilizing space probes that are to land. But the long survival times' of micro-organisms could make a danger of unsterilized space vehicles abandoned in orbit, since these could still be flying around in the solar system after centuries.

## The threat to life on Earth

We have learned now that we cannot regard this planet as being fenced in and a secure abiding-place for Man: we can never anticipate the unseen good or evil that may come upon us suddenly out of space. (H. G. Wells, *The War of the Worlds*).

This quotation from Wells's classic science fiction story, is the first appearance in literature of the theme of the 'threat to earth', since developed by many writers. Wells imagined the Martians as hostile intelligent beings (the first such in science fiction). Their invasion of the earth was defeated by terrestrial bacteria. Mars, said Wells, had no micro-organisms, and the Martians had no immunity to the most harmless of our microbes. We would now say that any world with life should at least have micro-organisms, so that Wells's story was muddled in this respect. Yet he did introduce the idea that harmless creatures could be the most potent of weapons against an alien life form. 'By the toll of a billion deaths', he wrote, 'man has bought the birthright of the earth'. But not, we may add, the birthright to other planets.

*Bacterial spores and other micro-organisms placed in this apparatus, which simulates the near vacuum of space, were all dead within thirty days. This and similar experiments suggest that it is extremely unlikely that micro-organisms could survive transfer from one planet to another by natural processes*

Our difficulty is that we cannot make any rational estimate of the dangers of bringing exobionts back to earth. To do this we would need wide biological knowledge based upon different life forms, and this is just what we lack. We can only extrapolate from our earthly experiences. It is very probable that organisms which have evolved on another planet would be unsuited to life on the earth. Certainly we need only consider carbon-based life with water as a solvent. Yet wherever a sizeable land mass has been long isolated from the rest of the world and contact has then been re-established, there has been disturbance resulting from invasion by new organisms. Examples of this are well-known: the rapid spread of the rabbit and the prickly pear in Australia, the Colorado beetle in Europe, the cornborer in America. The devastation of European and Australian rabbits caused by the myxoma virus from America is a striking recent example. It is of course true that only a few introduced forms establish themselves in their new habitat, and still fewer become pests, but the potential danger is undiminished.

We have few clues about what might happen if we were to introduce to the earth organisms of a different evolutionary background. They might be quickly destroyed by terrestrial organisms; on the other hand, they might be wholly indigestible. There is evidence that the range of foreign substances to which man and higher animals can become immune (an important aspect of disease resistance) is quite limited. It may be that we could not be immunized against a disease from another world, and we would not have much time to find antibiotics that were effective against it.

The earth could also be vulnerable in other unsuspected ways. J. J. Connington in his novel *Nordenholt's Million* tells the story of a new bacterium that denitrifies the soil, liberating all the fixed nitrogen that plants depend upon for their growth, with the consequent failure of agriculture throughout the world. Lederberg has noted that some micro-organisms produce poisonous substances as a by-product of their growth. For example, the 'red tides', caused by the vast multiplication of one-celled algae called dinoflagellates, which produce a powerful toxin, are responsible for the death of great numbers of fish, and of sea birds that eat the poisoned fish. Just occasionally, too, human beings are poisoned by shellfish such as mussels which absorb the poison. Exobionts might conceivably poison our crop plants in a similar way.

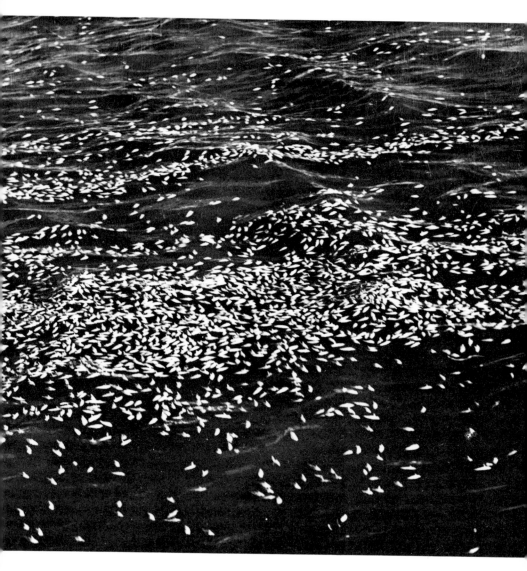

Another topical point concerns viruses. We are now beginning to understand how to put together the basic chemical compounds involved in life, and we are not far off being able to synthesize a virus. Viruses are on the border-line of the living and non-living, but they have the ability to deflect the mechanisms of

*Dead fish, victims of a 'red-tide' plague, floating on the swell near the Florida coast*

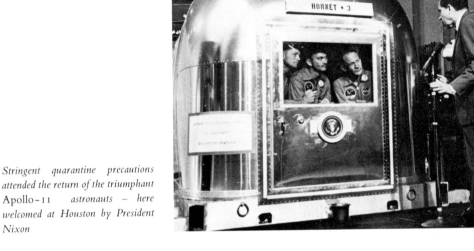

*Stringent quarantine precautions attended the return of the triumphant Apollo-11 astronauts – here welcomed at Houston by President Nixon*

cells toward their own ends, and to reproduce themselves with the aid of the machinery of the cells upon which they are parasitic. Almost incidentally, one might say, they harm the cells and cause disease. Viruses are in essence self-replicating molecules. It may be that other molecules can act like viruses, and recent work on the agent of the sheep disease called scrapie points this way. If so, there may be molecules on other planets that are not living organisms in any familar sense (and which might be extremely difficult to detect), yet which could behave like viruses in suitable hosts. Exobionts could show unusual powers of resistance to disinfecting agents. Those based on carbon could be destroyed of course by heat or oxidizing chemicals because of breakdown of carbon-to-carbon bonds. But if they were very hardy they would soon be distributed around the earth. Most microbes do not survive long in the air or in water, but transport across the oceans by winds and water is quite rapid.

We need not consider too seriously the dangers from the more exotic forms of life imagined by science fiction writers. H. G. Wells' Martians have been mentioned, and the SF aficionado could add a long list of gruesome examples invented by other writers. Again, we would be less concerned with the dangers to space explorers from mysterious diseases on far-flung planets; it is the protection of the earth which most concerns us. Fortunately the United States at least is well aware of the problem, and elaborate quarantine facilities are available at Houston, Texas. When the first men on the moon, the crew of *Apollo* 11, splashed down in the Pacific in July 1969 they were immediately quarantined. The lunar surface samples that they brought back with them were quarantined too, and sterilized before release. If contamination had been accidentally introduced on to the earth the technical problems involved in eliminating it might have been considerable. Our safest course would be to carry out precautionary experiments in space stations to test the effect of planetary material on small biological systems—soil, grass, small animals, etc.—in a sealed chamber. Experiments of this kind were carried out at Houston on the first samples to be returned from the moon—with negative results.

*A fragment of the first moon-rock sample to reach the earth is carefully examined under completely sterile conditions*

**187**

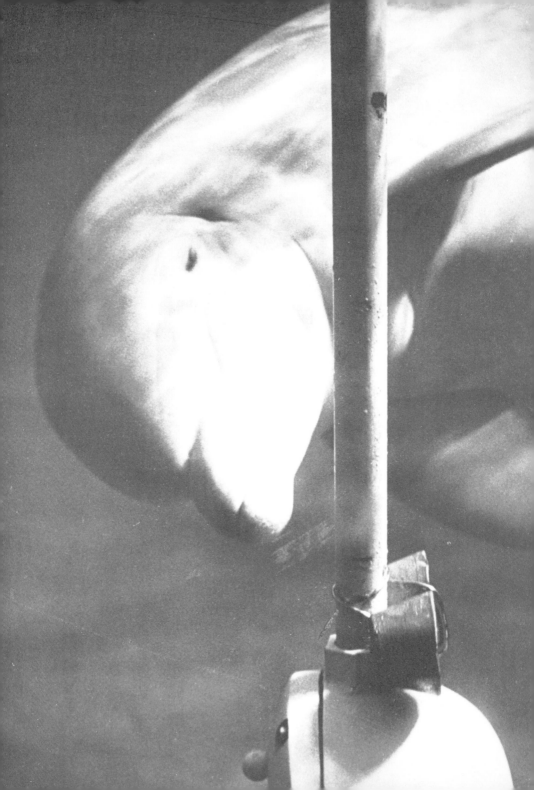

If we cannot journey to the distant stars, if the vast depths of space are never traversed by man, we may yet still hope to communicate with other worlds. If there are any intelligent beings within a reasonable distance of the earth it is possible that they are already trying to communicate with us. The idea of communicating with other planets is not new. In the last century the mathematician Gauss proposed the planting of a right-angled triangle of forest in Siberia as a sign that intelligent life exists on the earth. Sporadic suggestions of this kind were made over the years, but in 1959 G. Cocconi and Philip Morrison considered in some detail whether communication would be possible between civilizations belonging to different star systems.

## The Order of the Dolphin

In 1961 the Space Science Board of the U.S. National Academy of Sciences sponsored a meeting on 'Intelligent Extraterrestrial Life' at Green Bank, West Virginia, and since one speaker, J. C. Lilly, had been making a special study of the 'language' of dolphins, the meeting lightheartedly named itself 'The Order of the Dolphin'. The study of dolphin language is obviously important, because understanding these relatively intelligent mammals would clearly help us to understand the messages of other intelligent beings.

*Some porpoises and dolphins are no heavier than men, yet have larger and more extensively convoluted brains. It has been suggested that these aquatic mammals may prove to have a speech pattern as complicated and as intelligible as man's. Opposite: a porpoise examines an underwater loudspeaker and microphone*

If we assume that life does exist elsewhere in the universe, then it would seem reasonable that some forms of it have developed a technology at least as advanced as ours, and probably far more advanced. After all, we have only had technology for a short time. Other worlds may have had it for aeons. It is of course true that intelligent beings may not always be technological or wish to communicate with other worlds. But if there are other civilizations trying to talk across space, we can probably safely assume that they have language capable of abstract concepts, mathematics, and light and radio telescopes.

The most likely places for such civilizations would be on planets of stars like our sun, where life would be most likely to arise. Of the twelve stars nearest to the sun, only one—Epsilon Eridani—is a candidate. The nearest star of all, Proxima Centauri, is only 4·25 light-years away, but it is a member of the triple system of Alpha Centauri, and such a system is unlikely to have a suitable planet. The next, Barnard's Star, is very dim and cool, though it does have a planet. Likewise Wolf 359, Luyten 726-8 (a binary) and

*A star map locating the positions of Epsilon Eridani ( $\epsilon$ ) and Tau Ceti ( $\tau$ )—two nearby stars which may possibly be attended by life-bearing planets*

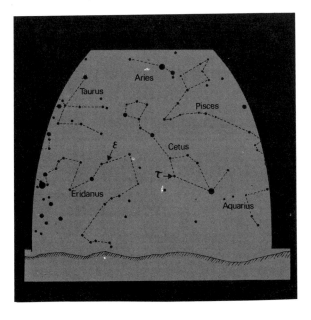

| Star system | Number in system | Colour | Distance (light-years) |
|---|---|---|---|
| **The sun** | 1 | yellow | 0 |
| **Alpha Centauri** | 3 | yellow, red, red | 4·3 |
| **Barnard's Star** | 1 | red | 6·0 |
| **Wolf 359** | 1 | red | 7·7 |
| **Lutyen 726-8** | 2 | red, red | 7·9 |
| **Lalande 21185** | 1 | red | 8·2 |
| **Sirius** | 2 | white, white dwarf | 8·7 |
| **Ross 154** | 1 | red | 9·3 |
| **Ross 248** | 1 | red | 10·3 |
| **Epsilon Eridani** | 1 | yellow | 10·8 |
| **Ross 128** | 1 | red | 10·9 |
| **61 Cygni** | 2 | red, red | 11·1 |
| **Lutyen 789-6** | 1 | red | 11·2 |
| **Epsilon Indi** | 1 | yellow | 11·3 |
| **Tau Ceti** | 1 | yellow | 12·2 |

*A table of nearby star systems indicating the number of stars in the system, the colour of each star, and distance (in light years) from the sun. A star's colour is a rough guide to its surface temperature: red stars are relatively cool, yellow stars warm, and white and blue stars extremely hot*

Lalande 21185 are dim stars. Then comes Sirius, the brightest star in the sky, with its dwarf companion, but it is too young a star to be suitable, as well as being a binary. We must also pass over the dim stars Ross 154 and Ross 348. Then at last we reach Epsilon Eridani, a single, fairly old and fairly warm star like our sun. Like the sun it rotates slowly, which is a sign that it may have planets. A little further away is Epsilon Indi at 11·3 light-years, another possibility, while at 12·2 light years is Tau Ceti. These last three are the likely candidates among the hundred or so stars within 20 light-years of us, and Tau Ceti is a star very similar indeed to the sun.

## Chances of making contact

There have been attempts to calculate the number of civilizations that are communicating with others in our galaxy. These follow the same sort of reasoning we used in estimating the number of planets supporting life, and are subject to the same deficiencies. In Chapter 5 we made an estimate of the number of life-bearing planets in our own galaxy, and concluded that it was between about 800-million and 200,000, with 12-million as the most probable figure. To extend this to the number of communicating civilizations we need to add some more terms to the formula used there. We need an estimate of the fraction of living systems that have developed intelligent life $(f_i)$, and also the fraction of these that wish to communicate $(f_c)$. These are usually taken as being quite high, almost 1·0. However, this is not at all sure. It may be that intelligence evolves rather seldom, and that non-technological civilizations predominate. We might advisedly prefer $f_c$ and $f_i$ both to be 0·1. But the biggest uncertainty comes with our last symbol, $L$, which is the fraction of the lifetime of the planet that the civilization actually broadcasts. This is anyone's guess. Intelligent beings might get bored after ten years of transmitting and listening, or they might keep it up for a million years. If the age of the average life-bearing planet is taken to be around 3,000-million years, this means that suitable planets may spend anything from, say, a thousandth to a hundred-millionth of their lifetime actually communicating. The value of $L$ is therefore critical.

It has also been suggested that technological civilizations are very likely to destroy themselves with their own inventions within a relatively short time. And quite apart from warfare there is the danger that it might be feasible, in the course of experimental research, to trigger unwittingly the annihilation of matter on a large scale, and thus blow up an entire planet by accident. Moreover, as Sagan notes, natural stellar explosions may be frequent enough to produce lethal bursts of radiation in most galaxies at frequent intervals, and thus kill any life soon after it arises. The worst feature of our estimate for $L$ is that it might be very short (for civilizations that destroy themselves) or else very long (for the few that do not). We should therefore perhaps choose a value like 1:100,000, but at the same time allow for a considerable margin of error. We may then obtain our estimate for the number of civilizations that are transmitting in our galaxy by multiplying the estimates in Chapter 5 of the number of life-bearing planets by $f_i \times f_c \times L$. Thus we obtain $1 \cdot 09 \times 10^{-3}$ and $1 \cdot 35 \times 10^3$ as our lower and upper limits with $1 \cdot 2$ as the most likely number. Thus, in round figures, there might be anything from a thousand such civilizations to a chance of only one in a thousand that there is even one of them. Again we see the difficulty of making any realistic estimate.

Over what distances could we hope to communicate? We could not communicate with other galaxies, because of the enormous lengths of time it would take to receive a reply. No suitable source has been suggested for the vast power required for intergalactic transmission. But for nearby stars the position is different. Calculations suggest that the limit of reception with beamed radio transmission and dish aerials is about 100 light-years. There are about 400 stars like the sun within this distance, and the energy required to transmit to them would not be excessive. If the transmissions were not beamed their range would of course be much less. One might expect, therefore, that civilizations would beam transmissions toward likely stars—towards those which had the best chance of being associated with life-bearing planets. Such messages may well be beamed toward us right now, since to other astronomers our sun would be a likely candidate. If laser light beams were used the range would be shorter, but visible light has a far greater capacity for carrying messages than radio waves. Very long messages could be sent in a very brief period of transmission; visible light, if modulated rapidly, could in only one second easily carry a million-million bits of information—equivalent to the contents of a large library. At present we cannot read such rapidly modulated signals, but they would have tempting advantages for the sender, since it could take years to send a comparable amount of information by radio. It is possible, however, that these considerations are

*Only a beamed radio transmission directed towards a specific star system within a radius of 100 light-years would have a significant chance of detection by an alien civilization. Signals broadcast to the universe at large would be extremely difficult to detect at distances of more than a few light-years*

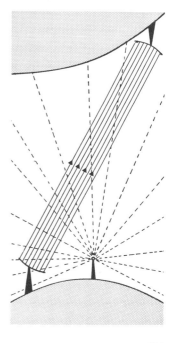

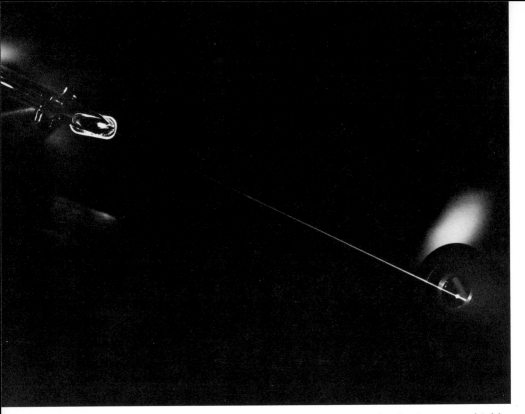

*Potentially far more efficient carriers of information than radio waves are the very much shorter waves of the laser. Laser light can be beamed with great precision and minimal loss of intensity over many millions of miles*

*Opposite: the world's largest equatorially-mounted radio telescope, sited near Greenbank, West Virginia. Its dish reflector, designed to amplify radio emissions from deep space, is 140 feet in diameter*

of less consequence to technologies more highly advanced than our own, which might have large energy sources at their disposal and a great deal of patience. It is also possible, as R. N. Bracewell has pointed out, that such civilizations might have sent automated space probes to orbit nearby stars, communicating with their base by a system of relay stations. Such probes would await the awakening of civilizations on planets and respond to signals from them.

The best wavelength to listen on has aroused especial interest. Most of the radio noise in the universe is due to the 21-centimetre-wavelength radiation emitted by hydrogen atoms in interstellar gas clouds, and, as Philip Morrison has pointed out, an intelligent being would assume that this wavelength was known to his audience. He would transmit at a frequency near this, or possibly some simple multiple of it. The argument is similar to that used if one were asked to

meet a friend in London, but neither of you knew where. Obvious places would be Piccadilly Circus or Trafalgar Square. The 21-cm. line is the Piccadilly Circus of the radio waves. It is also in the region where background noise is fairly low.

Lasers using visible light have received less attention. Because the light beam would be obscured by the light from the planet's sun, it might be tuned to a spectral absorption line of this star to allow better detection against the star's spectrum. There have even been suggestions that intelligent beings might modulate entire stars, and the recent discovery of pulsars raised the question of whether these pulsating radio stars were sending messages. No evidence for this has been obtained, however, and it is likely that pulsars are neutron stars, consisting of extraordinarily dense matter, that are rotating swiftly and in some way emitting both pulses of radio waves and light. The pulsar in the Crab Nebula has recently been shown to emit flashes of light at the same rate as the radio pulses.

The radio pulses vary in strength (although they are very regularly spaced) but this appears to be due to 'twinkling' because of erratic motion in interstellar gas clouds. Although pulsars would be ideal for communicating across great distances, because their intense transmissions would not need to be beamed, we have no idea how they could be modulated. If this were possible it seems likely that it would be easier to modulate some particular frequency than to suppress one pulse and not the next over all frequencies. If so, it may be worth examining the fine details of the pulses, and the ratios of intensities at different wavelengths, which might be less affected by twinkling than the absolute intensities.

## Project Ozma

In 1960 Frank Drake launched Project Ozma at the National Radio Astronomy Center at Green Bank, West Virginia. (Ozma is the name of the Princess in Frank Baum's children's book *The Wizard of Oz*.) Drake listened on an 85-foot radio telescope to the stars Tau Ceti and Epsilon Eridani. The noises were recorded and later analysed statistically to see whether there was any evidence of a regular or coherent pattern that might conceivably be a message. No convincing evidence was found. The Russians have also done similar work, though less is known about it. In 1964 they reported signals from 61 Cygni (11·1 light years away, but a binary star) which were said to be messages, but this report is regarded with scepticism, and no recent results have been announced. The Russians may have picked up weak pulses from a pulsar which gave the impression of artificial signals.

What sorts of message might come from other worlds? D. C. Holmes points out that there would be three kinds of signal: local radio traffic, beamed transmissions to regular correspondents, and messages intended for anyone who could receive them. The first two would be too faint, and would be unlikely to be pointed toward us. We would only expect, then, to hear the third. These might be beamed to likely stars at various times, or be unbeamed if their power was very great. It would be hard to believe that communicating beings were not familiar with mathematics, the speed of light, atomic structure and so on, so that series of numbers or the ratio of the mass of the electron to the proton would be the sort of thing we might expect. If we heard the following sequence of numbers: 1, 2, 3, 5, 7, 11, 13, 17 . . ., we would know at once that this was an intelligent communication; thunderstorms and other natural phenomena do not generate the prime numbers. Again, a number like 3·14159 would be recognizable as $\pi$ throughout the universe.

Another form of message would be a television picture, for we would expect intelligent beings to have pictorial ability. The recipient would not know the number of lines or the number of dots in a line, so one might choose prime numbers for both, for example 19 and 29. One frame of the television picture would consist of a series of exactly 551 dots and dashes. This series would be repeated many times. The

*A notional television picture of 19 lines, each consisting of 29 signals. It evidently shows a being who, as the Pythagorean diagram indicates, is intelligent and numerate. He is pointing to what must be the innermost of three planets, presumably his home. The marker to the right probably indicates his height, and the symbol 1011 may be a binary number. If so, it probably reads towards the left and stands for 13. The being's height is perhaps equivalent to 13 wavelengths of the frequency of transmission*

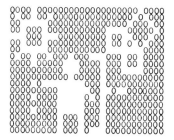

number would suggest to the hearer that he should arrange the signals either as 19 rows of 29 or as 29 rows of 19, and he would try both. After the 1961 conference at Green Bank, Frank Drake sent such a message to the participants for them to decipher (which most of them did), and several entertaining pictures of this sort have been published. There is perhaps a tendency for the sender to read more into the message than the receiver could be expected to see. Among other clues, the received signal might show a Doppler effect over a period of time as the planet of the sender rotated round its sun, and from this the length of year and perhaps other astronomical quantities could be calculated. For example, if we identified the spectral type of the star, we could estimate the temperature on the planet.

The signalling of messages can take two forms; it can use complex waveshapes, like speech, or it can involve on-or-off pulses, like the Morse code. The first is less easy to decipher against a noisy background, so one would expect interstellar communication to be in pulses, and perhaps in dots and dashes, which have some advantages. One would not then need to deduce the arrangement of a television picture for example, because pauses and double pauses would make this quite plain. But if a code of pulses is used it is immediately apparent that except for short series of numbers (which might consist of 1, 2, 3, 4 and more dots) one needs some convention to represent a unitary symbol, such as a letter or a digit. This convention is of course found in a television frame in a rather different form. If intelligibility is a prime requisite the symbols would best consist of a fixed number of pulses forming a group. The holes used on teleprinter tape are an example of this. Thus one can use o · · · · · o for 1, and · o · · · · o for the letter A. Numbers are now expressed, not in a string of dots, but in the familiar way. It is very likely that the base of the system would not be 10; other beings might not have ten fingers to set the base of their arithmetic. It might be set instead to a base of 8. The groups might be say 8 or 16 pulses, depending on the number of

symbols (probably rather numerous). There would need also to be spacers, and the whole thing would be very like a computer code. Certain positions in the group would doubtless be reserved for special classes of symbol (letters, numerals) as is often done with computers. The pulses would probably be amplitude modulated (whether frequency modulation would be better has received little attention). A startling proposal, by L. C. Edie, is that the long chain organic molecules in certain meteorites, like the Orgeuil meteorite, were placed there by a distant civilization, and

*The huge radio telescope at Arecibo, Puerto Rico. A light aluminium mesh, suspended in a valley, forms the vast reflector – 1,000 feet across*

launched toward us. Could these molecules contain messages, he asks, in the way that proteins can?

The detection of coherence or pattern in a noisy signal is in theory quite simple. We should, however remember that one cannot test for a pattern that one has not thought of. Highly intelligent beings might employ patterns that were unintelligible to us. However, it is clear that the occurrence of groups and sets of groups would not be random; thus the + sign will nearly always lie between two numerals or two algebraic letters. If special positions in the group have special meanings this would soon be apparent.

## Interstellar languages

Cocconi and Morrison, discussing the nature of signals that might be received from other civilizations, have argued that they would each probably consist of three parts. There would be short sections to draw attention (sequences of prime numbers, for example). There would be language lessons, that would teach the meaning of symbols and the like. Then, because of the long time before a reply could be received from another world, it would seem rational to send a long section of information of the kind found in an encyclopedia. The advantage of this scheme would be that there would be a good chance that something of the message could be read wherever the hearer happened to start receiving it. As Morrison put it, we need a new special study: 'anticryptography', the science of making codes as easy as possible to decipher. The signals to draw attention might of course be ones that carried no message. Thus it has been suggested that a technology might dump a few hundred tons of an artificial radioactive element into a star, so that this element produced a spectroscopic line which would be recognized as artificially induced. Such a star would invite closer attention.

Presumably a special and simple language would be used. H. Freudenthal has developed a language, 'Lincos', for just such a purpose, and outlines the way different concepts would be transmitted. He notes that an intelligent being learns from a few examples. Thus if one sees $5 = 3 \, \P \, 2$, $7 = 3 \, \P \, 4$, $19 = 13 \, \P \, 6$, it is soon obvious that the symbol $\P$ stands for $+$. In this way one can convey the mathematical symbols, concepts like 'not', 'prime', 'because', 'nearly', and so on, and this should allow finally the transmission of nouns, like 'man', and verbs like 'give'.

A neat example of a message using simple arithmetic was given by Ivan Bell in *The Japan Times* and reprinted in *Scientific American*, August 1965. A short part is given below. It uses the alphabet, and first by illustration defines the numerals 1 to 9 as the letters A to I. The reader may like to try deciphering the meaning of some other letters from the examples below.

AKALB    AKAKALC    IMGLB    CKNLC    EMELN    JLAN
JKALAA    EPBLJ    JQBLE    AQJLU    ULWA    GIWIHYHN
ZYCWADAEI

Such messages may nevertheless be difficult for us to decipher if superior intelligences send them, because they may contain many abstract and difficult concepts, and be more dependent on context than we might think. Animals use 'analogue' language (a dog's growl means the same thing whenever uttered) while we use a more 'digital' language, where meaning depends on position, just as the 7 in 179 and in 0·73 do not mean the same thing. Superior intelligence would be indicated by proofs of mathematical theorems which we have not solved, though we might not be able to understand the proofs.

## Flying saucers

Some have puzzled over why, if superior beings exist elsewhere, they have not communicated with us or visited us. At least one might expect them to have left records of their visits. It may be that they have, but that we have not yet discovered them. Some believe that there have been such visits, and that ancient legends of gods and heroes refer to them; others that visits were made millions of years ago before man evolved. There are also many who believe that

*Key to the symbols used in the extracts from a coded message quoted on this page:*

| A | B | C | D | E | F | G | H | I | J |
|---|---|---|---|---|---|---|---|---|---|
| 1 | 2 | 3 | 4 | 5 | 6 | 7 | 8 | 9 | 10 |

| K | L | M | N | P | Q |
|---|---|---|---|---|---|
| + | = | − | o | × | ÷ |

'Flying Saucers', more coldly referred to as Unidentified Flying Objects (UFOs), are spaceships of such beings, who are able to keep well hidden if they desire. Although investigations leave a small residue of UFOs unexplained, there seems to be no plausible evidence that they are alien spaceships, and Sagan notes significantly that there have been no reports of Flying Saucers from astronomers who watch the sky for long periods and photograph it extensively. Our radio signals, however, have been spreading into space for the last forty or fifty years, so if they are heard elsewhere we may be visited soon!

Most people who believe in Flying Saucers believe that their pilots are well disposed towards us, but perhaps we should not count on the beneficence of aliens. It is an interesting question whether intelligence leads to charity toward other intelligences. It could be argued, I suppose, that intelligent beings must be passionately curious or their intelligence regresses, and being curious they will always preserve other life forms lest they destroy the objects of their own intellectual passions. Such beings may have such a command of technology that they now value other sentient beings above all else. What indeed would they want of us? Not food or slaves. Perhaps some peculiar talent we have? Yet perhaps they would only wish to stamp out all opposition. Again, what is valuable to us may be dross to another civilization, or vice versa. H. B. Fyfe has a neat story, 'In Value Deceived', of how two men acquire a transmuter of elements from two aliens, who in turn acquire a box of lettuce from the men, and each party feels guiltily that they have stolen a treasure which in reality the other does not value at all.

However, if the vast distances of space are an impassable barrier to travel, then only messages can cross it. We might learn a great deal from other civilizations —not least, as Hoyle points out, the way to control our own society in the face of our aggressive instincts and excessive birth rate. The knowledge might liberate us from the wheel of the rise and fall of civiliza-

tions. We would expect to learn much, not only in the field of technology, but also in art and philosophy. Great advances in medical knowledge might be achieved and much suffering prevented. It has been argued that we need not fear other beings if interstellar travel is impossible and knowledge is always beneficial. But others see a threat to our existence in knowledge itself. An innocent-sounding experiment which we did not understand, prescribed by an alien civilization, might produce a nuclear explosion. Humans may be subject to control by superior information, just as we are able to control a flock of sheep through a sheepdog. Anthropologists are familiar with the great strains that may tear apart primitive societies when new ideas are brought in, and with the consequent appearance of strange obsessions like the New Guinea cargo cults.

*A chilling scene from a film adaptation of the theme of H. G. Wells's* War of the Worlds. *The United States Air Force has recently closed its files after 21 fruitless years of investigating reports of 'Unidentified Flying Objects', but should a future UFO ever prove to be an alien space craft, one thing is certain: it will not be piloted by Martians*

The psychologist C. G. Jung observed that contact with superior beings might be shattering to us; to find ourselves no more a match for them intellectually than our pets are for us, to find all our aspirations outmoded, might leave us completely demoralized.

Again, Fred Hoyle, in 'Ossian's Ride', describes how an alien intelligence is transmitted to the earth in a train of waves to be reconstituted as a human being. Can we imagine a self-organizing message? Could a train of waves coax a nucleic acid chain into being from its constituent atoms? Such a concept would open the way for space travel in all but name.

At present we can do little but listen. We have neither the energy resources nor the theoretical knowledge of biology and language to beam transmissions toward stars with much hope of profit. We have seen that there may be few or many civilizations in our galaxy, and that the chances of there being one close to us are not very great, even on our more optimistic estimates. But Cocconi and Morrison, in pointing out the importance, practical and philosophical, of detecting interstellar communication, added this: 'The possibility of success is difficult to estimate; but if we never search, the chance of success is zero'.

# GLOSSARY

**abiogenesis:** the production of complex carbon compounds in the absence of life.

**absolute zero:** the temperature at which movement of molecules ceases, i.e. $-273 \cdot 16°C$.

**alleles:** alternative forms of the same gene (q.v.).

**anaerobe:** an organism that can live without free oxygen, and consequently does not need contact with air (either as a gas, or in solution in water).

**antimatter:** matter composed of subatomic particles with opposite charge to those composing ordinary matter. Matter and antimatter annihilate each other on contact. It is possible that some galaxies consist of antimatter.

**apparent brightness:** the brightness of a star as seen from the earth. It depends both on the star's distance from us and its 'absolute brightness', or luminosity.

**atmosphere:** the gaseous mantle of a heavenly body. It is also used as a unit of pressure, being $14 \cdot 7$ lb. per square inch, the pressure of the earth's atmosphere at sea level.

**black-body radiation:** the idealized radiation that would be emitted by a body that absorbed all the radiation that it received. It has a special spectral composition.

**carbonaceous chondrite:** a stony meteorite containing large amounts of carbon compounds.

**catalyst:** a substance that increases the rate of a chemical reaction without being used up itself.

**chemoautotrophs:** bacteria that can use certain simple chemical compounds (e.g. sulphur) as their sole source of energy and obtain their carbon and hydrogen from carbon dioxide and water.

**chromatography:** a method of chemical analysis in which the sample is moved through a porous material by a current of liquid or gas. Different chemical substances move at different rates, so the mixture is separated into its components.

**chromosomes:** the rod-shaped bodies that carry the genes (q.v.) and which form most of the cell nucleus.

**critical temperature:** the temperature above which it is impossible to liquefy a gas by increasing the pressure.

**deoxyribonucleic acid (DNA):** one of the two forms of nucleic acid (q.v.). DNA contains the sugar deoxyribose, phosphate groups and the bases adenine, cytosine, guanine and thymine, and it is the chemical that carries the genetic message of the genes (q.v.).

**Doppler effect:** the fall in frequency of a wave when the source is receding from the recipient, or the rise in frequency when the source is approaching. Familiar as the change in pitch of a train whistle as it passes by.

**enzyme:** a protein (q.v.) that is a catalyst (q.v.).

**eobionts:** 'dawn organisms', the earliest living creatures, that arose at least 3,500 million years ago.

**escape velocity:** the speed that a rocket must attain (in the absence of atmospheric resistance) to escape from the gravitational pull of a planet.

**eucaryotes:** organisms whose cells possess a cell nucleus, containing the chromosomes, surrounded by cytoplasm containing mitochondria (and other complex bodies). They comprise all organisms except bacteria and blue-green algae.

**exobionts:** organisms of worlds other than the earth (and not terrestrial in origin).

**gene:** the bearer of an inherited characteristic. The genes are composed of deoxyribonucleic acid (q.v.) and are arranged along the chromosomes (q.v.).

**ionization:** the act of forming ions, or electrically charged molecules. 'Ionizing radiation' is radiation that produces many ions in passing through water (e.g. X-rays, gamma rays, fast electrons or short ultra violet rays). An 'ionizing solvent' is a liquid in which dissolved salts dissociate into ions.

**isotope:** a given chemical element can consist of several kinds of atoms, isotopes, which differ in their weight because of different numbers of neutrons in the atomic nucleus.

**laser:** a device for producing an intense beam of light in which all the waves are in phase (in step together). As a consequence the beam is almost parallel even over enormous distances.

**light-year:** the distance travelled by light in one year, 5·88 million million miles.

**Main Sequence**: the principal series of the evolution of stars, with a characteristic position in a graph plotting star temperature against absolute brightness. Our sun belongs to the Main Sequence.

**megacycles**: a unit of frequency of radio waves, being the number of millions of waves per second. Megacycles are related inversely to the wavelength through the speed of light (186,000 miles/second, 300,000 kilometres/sec). Thus 1 mc. = a wavelength of 300 metres, and 100 mc. = a wavelength of 3 metres.

**micro-organisms**: these are all small organisms of simple structure. It is usual to include in them the bacteria and blue-green algae (which are procaryotes, q.v.) and the protozoa, the smaller fungi and the smaller forms of other algae (which are eucaryotes, q.v.).

**mutation**: a sudden change in a gene which is then inherited in the new form. Many mutations consist of alterations in nucleotide bases, and these cause alterations in protein structure.

**nebulae**: clouds of stars, dust and gas in interstellar space. Some are close to us, like dust clouds in the Milky Way, but others, the galaxies, are great systems of stars like the Milky Way itself. Galaxies are sometimes called spiral nebulae.

**nucleic acids**: polymers (q.v.) formed of nucleotides. The nucleotides consist of sugar molecules, phosphate groups and nucleotide bases. There are two types of nucleic acid, *deoxyribonucleic acid* (q.v.) and *ribonucleic acid* containing ribose as the sugar, and the nucleotide bases adenine, cytosine, guanine and uracil.

**nucleus**: the *atomic nucleus* is the central part of an atom, composed of protons and neutrons, around which the electrons are orbiting. The *cell nucleus* is the central part of a eucaryote (q.v.) cell, containing the chromosomes (q.v.).

**parsec**: an astronomical unit of length, which equals 3.259 light-years. It is the distance of a star at which the radius of the earth's orbit subtends an angle of one second of parallax.

**pH**: a scale for measuring acidity and alkalinity. pH 0 is very strongly acidic, pH 7 is neutral and pH 14 is very strongly alkaline.

**photosynthesis**: the process by which plants obtain energy from sunlight, and convert carbon dioxide and water into complex organic compounds.

**photon**: a quantum (the smallest amount) of electromagnetic radiation, including radiation of short wavelength (e.g. gamma rays), medium wavelength (e.g. visible light) and long wavelength (e.g. radio waves).

**polymer**: a large chemical molecule formed by the joining together of many simpler molecules (monomers). Nucleic acids, proteins and plastics are examples of polymers.

**procaryote**: an organism whose cells do not contain a nucleus clearly separated from the cytoplasm, whose DNA forms a simple naked thread-like chromosome, and which lacks mitochondria. They comprise only the bacteria and blue-green algae.

**proteins**: polymers (q.v.) formed of amino acids, which with nucleic acids are the critical components of terrestrial organisms.

**psychozoa**: 'mind-life'. Terrestrial organisms that have high intelligence, man being the first of these to evolve.

**radioactivity**: the phenomenon of spontaneous breakdown of certain atomic nuclei to give atoms of other elements (e.g. the decay of the metal radium to the noble gas radon), accompanied by various forms of ionizing radiation.

**red shift**: the shifting of spectral lines of galaxies toward the red end of the spectrum. If interpreted as a Doppler effect (q.v.) it indicates that the galaxies are receding from us.

**Roentgen**: a unit of ionizing radiation in which X-ray dosage is measured.

**spectroscopy**: the study of spectra of electromagnetic radiation, including both emission (as by stars) and absorption (as by the atmosphere of a planet). Chemical composition can often be determined in great detail. Spectrometry is one aspect of spectroscopy.

**stratosphere**: the upper layer of the earth's atmosphere, above about 7 miles at the poles and 10 miles at the equator.

**supernova**: an exploding star, which becomes as bright as 10 million suns because of violent thermonuclear reactions within it.

**symbionts**: organisms that live intimately together to their mutual benefit. Lichens are symbiotic associations of fungi and algae.

**valency**: the number of chemical bonds that atoms of an element can form with other atoms.

**xenobionts**: a life form based on a different chemistry from that of terrestrial organisms.

# BIBLIOGRAPHY

## General

Shklovskii, I. S. and Sagan, C., *Intelligent Life in the Universe*. San Francisco, 1966.

Sullivan, W., *We Are Not Alone*. London, 1964.

Holmes, D. C., *The Search for Life on Other Worlds*. New York, 1966.

Shapley, H., *Of Stars and Men*. Boston, 1958.

Hoyle, F., *Of Men and Galaxies*. Seattle, 1964.

Firsoff, V. A., *Life Beyond the Earth*. London, 1963.

Mamikunian, G. and Briggs, M. H. (eds.), *Current Aspects of Exobiology*. Oxford, 1965.

Schneour, E. A. and Ottesen, E. A. (eds.), *Extraterrestrial Life: an Anthology and Bibliography*. Washington, D.C., 1966.

## 1 The Physical Universe

*The Space Encyclopaedia*, 2nd. edn. Horsham, Sussex, 1964.

Motz, L. and Duveen, A., *Essentials of Astronomy*. London, 1966.

Wyatt, S. P., *Principles of Astronomy*. Boston, 1964.

Gamow, G., *Mr. Tompkins in Paperback*. London, 1967.

Burhop, E. H. S., 'The 300 Ge V accelerator: the programme'. *Science Journal*, July 1967, 39-45.

Reynolds, J. H., 'The Ages of the Elements of the Solar System'. *Scientific American*, November 1960, 171-80.

## 2 Terrestrial Life

Anfinsen, C. R., *The Molecular Basis of Evolution*. New York, 1963.

Simpson, G. G., *This View of Life*. New York, 1965.

Lanham, U., *Origins of Modern Biology*. New York, 1968.

Abelson, P. H. 'Paleobiochemistry'. *Proceedings of the Fifth International Congress of Biochemistry*, Vol. 3, 52-68 (1963).

Keilin, D., 'The problem of anabiosis or latent life'. *Proceedings of the Royal Society of London*, Series B, Vol. 150, 149-91 (1959).

Sneath, P. H. A., 'The limits of life'. *Discovery*, Vol. 25, 20-4 (1964).

—'Longevity of micro-organisms'. *Nature*, Vol. 195, 643-6 (1962).

## 3 The Nature of Living Systems

Bernal, J. D., *The Origin of Life*. London, 1967.

Penrose, L. S., 'Self-reproducing machines'. *Scientific American*, June 1959, 105-11.

Perret, C. J., 'A new kinetic model of a growing bacterial population'. *Journal of General Microbiology*, Vol. 22, 589-617 (1960).

## 4 The Origin of Terrestrial Life

Oparin, A. I., *The Origin of Life*, 2nd. edn., translated by S. Morgulis. New York, 1953.

Oparin, A. I. (ed.), *The Origin of Life on the Earth: Reports on the International Symposium*. Moscow, 1957.

*Journal of the British Interplanetary Society*, Vol. 21, 2-112 (1968).

## 5 Life Beyond the Earth

Calvin, M., 'From Molecules to Mars'. *American Institute of Biological Sciences Bulletin*, October 1962, 29-44.

Quimby, F. H. (ed.), *Concepts for Detection of Extraterrestrial Life*. Washington, D.C., 1964.

Lovelock, J. E., 'A physical basis for life detection experiments'. *Nature*, Vol. 207, 568-70 (1965).

Michaux, C. M., *Handbook of the Physical Properties of the Planet Mars*. Washington, D.C., 1967.

Koenig, L. R., Murray, F. W., Michaux, C. M. and Hyatt, H. A., *Handbook of the Physical Properties of the Planet Venus.* Washington, D.C., 1967.

## 6 Alternative Biochemistries

Henderson, L. J., *The Fitness of the Environment.* New York, 1913; Boston, 1958.

Allison, A., 'Possible forms of life'. *Journal of the British Interplanetary Society*, Vol. 21, 48-51 (1968).

Barth-Wehrenalp, G. and Black, B. P., 'Inorganic polymers'. *Science*, Vol. 143, 627-633 (1963).

Firsoff, V. A., 'An ammonia-based life'. *Discovery*, Vol. 23, 36-42 (1962).

Pirie, N. W., 'Germanium as a carbon analogue'. In Good, I. J. (ed.), *The Scientist Speculates.* New York, 1962.

## 7 Intelligent Life

Fink, D. G., *Computers and the Human Mind.* New York, 1966.

Haldane, J. B. S., *Possible Worlds and Other Essays.* London, 1927.

Good, I. J., 'Life outside the Earth'. *The Listener*, 3 June 1965, 815-7.

Williams, W. T., 'Computers as botanists'. *Nature*, Vol. 197, 1047-9 (1963).

Bitterman, M. E., 'The evolution of intelligence'. *Scientific American*, January 1965, 92-100.

Wiener, N., *The Human Use of Human Beings.* New York, 1950, 1967.

—*God and Golem, Inc.* Cambridge, Mass., 1964.

## 8 Space Travel

Clarke, A. C., *The Promise of Space.* London, 1969.

Pirie, N. W. (ed.), 'The Biology of Space Travel'. *Institute of Biology Symposium No. 10.* London, 1961.

N.A.S.A., *Spacecraft Sterilization Technology.* Washington, D.C., 1966.

Lederberg, J. and Cowie, D., 'Moondust'. *Science*, Vol. 127, 1473-5.

Parkes, A. S. and Smith, A. V., 'Transport of life in the frozen or dried state'. *British Medical Journal*, 16 May 1959, 1295-7.

## 9 Interstellar Communication

Cameron, A. G. W. (ed.), *Interstellar Communication.* New York, 1963.

Drake, F. D., 'How can we detect radio transmissions from distant planetary systems?' *Sky and Telescope*, Vol. 19, 140-3 (1960).

Gardiner, M., 'Thoughts on the task of communication with intelligent organisms on other worlds'. *Scientific American*, August 1965, 96-100.

Freudenthal, H., 'Lincos: design of a language for cosmic intercourse.' Part I. Amsterdam, 1960.

Cocconi, G. and Morrison, P., 'Searching for interstellar communication'. *Nature*, Vol. 184, 844-6 (1959).

## Science Fiction

Asimov, I., 'Not Final!' *Astounding Science Fiction*, October 1961.

Asimov, I., 'The Last Question'. In *Nine Tomorrows.* London 1959.

Capek, K. and Capek, J., *R.U.R. and The Insect Play*, 1923; translated by P. Selver. Oxford, 1961.

Connington, J. J., *Nordenholt's Million* (1923); reprinted, Harmondsworth, 1946.

Dickson, G. R., 'The Monkey Wrench'. In *More Penguin Science Fiction.* Harmondsworth, 1963.

Forster, E. M., 'The Machine Stops'. In *The Eternal Moment and Other Stories.* New York, 1928. Also in *Collected Short Stories.* Harmondsworth, 1954.

Fyfe, H. B., 'In Value Deceived'. *Astounding Science Fiction*, November 1950.

Hoyle, F., *The Black Cloud.* London, 1957.

—*Ossian's Ride.* London, 1967.

Lewis, C. S., *Out of the Silent Planet.* London, 1938.

Stapledon, O., *Last and First Men.* London, 1930, 1963.

—*Sirius.* London, 1944, 1964.

Wells, H. G., *The War of the Worlds.* London, 1898, 1967.

—*The First Men in the Moon.* London, 1901, 1964.

Wyndham, J., 'Meteor'. In *The Seeds of Time.* London, 1956.

# SOURCES OF ILLUSTRATIONS

*Page*

8 Earth photographed on lunar horizon by *Apollo* 11. United States Information Service (USIS)

10 The comet Ikeya-Seki, October 1965. USIS

11 Late medieval representation of the flat earth, heavens and celestial spheres. Ronan Picture Library

12 The universe according to Ptolemy (top) and Copernicus (bottom). Courtesy of Weinreb and Douwma Ltd, London

13 Isaac Newton: painting by William Blake. The Tate Gallery, London

14 200-inch Hale telescope. Mount Wilson and Palomar Observatories

15 Spiral Nebula in Ursa Major. Mount Wilson and Palomar Observatories

16 210-foot radio telescope at Parkes, New South Wales. Photo: E. McQuillan

17 Crab nebula. Mount Wilson and Palomar Observatories. © Copyright 1959 by California Institute of Technology and Carnegie Institution of Washington

18 Cygnus-A radio source. A. T. Moffett

Quasi-stellar radio source 3C273. Mount Wilson and Palomar Observatories

19 Individual pulses from CP0328. Courtesy of *Science Journal*, London

21 Great nebula in Orion. Mount Wilson and Palomar Observatories. © Copyright 1959 by California Institute of Technology and Carnegie Institution of Washington

23 Field-ion micrograph of a tungsten tip. Erwin Müller, Pennsylvania State University

25 Cork cells. Robert Hooke: *Micrographia*, 1665

26 Comparison of a human body with a house. Tobias Cohn 1708. Courtesy of the Wellcome Trustees, London

Francesco Redi. *Opere*, 1809

27 Pasteur's curved-neck flasks; Louis Pasteur as a young man. Institut Pasteur

28 Gregor Mendel. Radio Times Hulton Picture Library

29 Radiation-induced mutations in fruit flies (*Drosophila melanogaster*). Brookhaven National Laboratory, New York

Chromosomes. Courtesy of W. Beermann

30 Metabolic pathways. D. E. Nicholson

32 Model to show folding of the myoglobin molecule. *Sunday Times*. Photo: Ian Yeomans

Atomic model of the myoglobin molecule. H. C. Watson

34 Model of part of the DNA molecule. M. H. F. Wilkins, Medical Research Council Biophysics Unit, London

35 DNA replication. Thames & Hudson Archives

37 Polysomes from rabbit reticulocyte cells. Courtesy of Alexander Rich

38 Section through the lichen *Xanthoria*. Gene Cox

39 Nematode worms caught by a fungus. C. L. Duddington

A fly trapped on a sundew leaf. L. Hugh Newman

40 Edelweiss. Swiss National Tourist Office

41 Hot springs, New Zealand. D. Brazier

42 Giant saguaro cactus, Arizona. United States Travel Service

43 The lichen *Xanthoria parietina*. Heather Angel

45 Fermentation vat in a malt whisky distillery, Perthshire. Distillers Company Ltd

46 Astronaut Edward White in space. USIS

47 Herald reactor at Aldermaston. United Kingdom Atomic Energy Authority

49 Bristlecone pines, California. U.S. Forest Service

50 Electron microscope scanning photograph of *Bacillus polymyxa* spores. J. Murphy, L. Campbell and the Editor of ASM Publications

51 Cybermen from 'Dr Who' science-fiction adventure. BBC, London

*Saturn*-I launch, May 1965. USIS

52 Arctic lupin plants grown from 10,000-year-old seeds. Canadian Department of Agriculture, Research Branch

54 Sequoias in Yosemite National Park, California. National Park Service

56 Candle flame. Courtesy of Carl Zeiss

57 Growth spiral on silicon carbide crystal. Courtesy of S. Tolansky

Bovine cortex: two vessels surrounded by osteocytes. Andrew L. Bassett

60 Queen bee surrounded by workers. Colin Butler

62 Cellulose lamellae in walls of *Chaetomorpha melagonium*. Eva Frei and R. D. Preston

65 Crystal of rock salt. Chr. Belser Verlag, Stuttgart

Sodium chloride atomic model. Courtesy Crystal Structures Ltd

**211**

67 Bacteriophage (T4) infecting *Escherichia coli*. Oak Ridge National Laboratory

Human poliomyelitis virus, type II. University of California Virus Laboratory

68 *E. coli* bacteriophage (T2) releasing DNA molecule after osmotic shock. Department of Biochemistry, New York University School of Medicine

71 Double-hook units. From 'Self-Reproducing Machines' by L. R. Penrose. *Scientific American*, June 1959

72 Double layers in sections of proteinoid microspheres. Electron micrograph. Sydney W. Fox

74 Alexander Oparin. Novosti Press Agency

76 Stanley L. Miller and his 'origin of life' experiment. Courtesy of Stanley L. Miller

77 Ultra-violet generator and primordial synthesis apparatus. Ames Research Center, National Aeronautics and Space Administration (NASA)

79 Proteinoid droplets. Institute of Molecular Evolution, University of Miami

81 Algal fossil from Gunflint rock, Ontario. Courtesy of E. S. Barghoorn

82 Blue-green alga: *Merismopedia glauca*. Photo: P. Echlin

Cross-section of a chloroplast. Courtesy of K. Mühlethaler

83 Part of bat pancreatic cell. J. Brachet

84 Blue-green alga: *Chroococcus turgidus*. Photo: Paul Curtis

85 The green alga: *Glaucocystis nostochinearum*. Photo: Paul Curtis

88 *Apollo*-8 launch, December 1968. USIS

91 Transit of Mercury, November 1907. Yerkes Observatory Photograph

92 Mercury probe planned for 1975. Messerschmitt-Bölkow-Blohm, Munich

93 Phases of Venus. Lowell Observatory Photograph

94 Capsule of *Venera* 4. Novosti Press Agency

95 Radar map of part of the surface of Venus. USIS

96 Earth photographed from *Apollo* 11, July 1969. USIS

97 Sunset viewed through the earth's atmosphere, photographed from *Gemini* 4. USIS

98 Seasonal changes on Mars; March, May, June, and July. Lowell Observatory Photograph

99 Edge of south polar ice cap on Mars, photographed by *Mariner* 7. USIS

101 Mars. Lick Observatory

102 Drawing of Mars by Giovanni Schiaparelli, 1879. Osservatorio Astronomico di Brera, Milan

Drawing of Mars by Percival Lowell, 1894. Ronan Picture Library

103 *Mariner*-6 photograph of Martian craters south of the equator. USIS

104 Jupiter. Kitt Peak National Observatory

Electric discharge process near frozen surface in a Jupiter simulator. Ames Research Center, NASA

105 Saturn. Lick Observatory

107 Uranus and its satellites. Yerkes Observatory Photograph

Pluto, showing motion. Mount Wilson and Palomar Observatories

108 Craters on far side of the moon, photographed from *Apollo* 10. USIS

109 Edwin Aldrin standing on the moon, July 1969. USIS

110 Four photographs of lunar rocks and dust, taken in July 1969. USIS

111 Moon photographed from *Apollo* 11. USIS

112 Lunar station of the future. Courtesy of *Science Journal*, London

113 Solar prominence. High Altitude Observatory, Boulder, Colorado

115 Organized body from Orgeuil meteorite. B. Nagy

117 Life zone of a solar system. Thames & Hudson Archives

119 Projected Mars research laboratory. USIS

120 Surveyor television camera. Hughes Aircraft Company

122 Abbreviated microscope. NASA

123 'Gulliver' life-detection device. NASA

124 The Multivator. NASA

127 Part of the southern Milky Way, photographed in red light. Courtesy of A. D. Code and T. E. Houck

130 Snow crystal. Supplied by the Meteorological Office Library

Diamond. De Beers Consolidated Mines Ltd, London

136 Spectrum of Saturn and Jupiter, showing ammonia bands. Mount Wilson and Palomar Observatories

140 Red blood corpuscles in human capillary. P-I Branemark

141 Flints. M. MacDonald

Silicon crystal grown from vapour condensation. IBM

145 A 'mutant' from the film 'This Island Earth'. Universal International

148 Blindfold porpoise. Globe Photos Inc. Photo: Dick Hewett

149 Multiple exposure photograph of the bat *Eptesicus fuscus* catching a mealworm. F. A. Webster

150 Head of Diamondback rattlesnake. Zoological Society of London

151 Venus fly trap (*Dionaca muscipula*), before and after catching a fly. Wilhelm Hoppe

153 Striped dolphins in 18-foot formation leap. Marineland of the Pacific Photograph

154 Chimpanzee and 'grape' shy. Michael Lyster

157 Solid logic modules. IBM

159 Computer programmed to play draughts. IBM

160 Cybermen from 'Dr Who' science fiction adventure. BBC, London

162 'Intelligent' hand-eye machine. Courtesy of John M. Snyder and *Science Journal*, London

163 Robot built by Captain Richards in 1928. Radio Times Hulton Picture Library

165 Computer 'Hal' from Stanley Rubrick's '2001'. MGM

166 The ascent stage of the *Apollo*-11 lunar excursion module leaving the moon, July 1969. USIS

167 Yuri Gagarin in the cabin of *Vostok* 1, 1961. Novosti Press Agency

168 Wide-angle lens view of *Apollo* 11 being launched. USIS

169 *Saturn* V and the *Apollo* spacecraft. Thames & Hudson Archives

170 Projected Mars electric propulsion vehicle. Courtesy of E. Stuhlinger

171 Ion engine in a vacuum chamber, and nuclear rocket engine under test in Nevada. USIS

172 Interior of *Apollo* command module. North American Aviation Inc.

173 Artist's conception of a manned orbiting space station. NASA

174 Effect of very low pressure on water. U.S. Air Force Photo

175 Volunteers in a life-support system designed for long-duration space missions. USIS

177 Artist's conception of a photonic rocket. Boeing Company

178 Deep-freeze compartments in space ship. From Stanley Rubrick's film '2001'. MGM

179 Time effects of the journey between Earth and Proxima Centauri. Thames & Hudson Archives

183 Scientist testing the survival of spores in a space simulator. USIS

185 Fish killed by the 'red tide', a population explosion of the dinoflagellate *Gymnodinium breve*. U.S. Bureau of Commercial Fisheries

186 *Apollo*-11 astronauts in quarantine in Houston, Texas. USIS

187 Lunar rock being examined. USIS

188 Porpoise examining loudspeaker. Marineland of Florida

194 Gas laser under test. Ferranti Ltd

195 140-foot radio telescope at Green Bank, West Virginia. USIS

199 Arecibo Observatory, Puerto Rico. USIS

203 Flying saucers from the Paramount film 'War of the Worlds'. Radio Times Hulton Picture Library

**213**

Abelson, P. H. 53
abiogenesis 75–8, 116, 205
absolute zero 53, 205
acids and alkalis 44
air and life 44
Aldrin, E. *109*
algae 38, *38*, 43–4, *81*, 96
  blue-green 42, 46, 81–4, *82*, *84–5*
  green *85*, 175
alleles 29, 205
Allison, A. C. 143
amino acids 31, 36, 53
ammonia 76, 105–7, 135
  liquid 135–8
  spectrum of *136*
anaerobes 44, 80, 205
Anders, W. 167
antimatter 176, 205
apparent brightness 15, 205
Archimedes 10
Aristotle 25
Armstrong, N. *109*, 167
Arrhenius, S. 74
Asimov, I. 163
asteroids 113–4, *114*
astronauts 8, 108, *109*, 173, 177
atmospheres *86*, 89–114, 205
atomic structure 23
ATP 32, 79, 121, 177
Avery, O. T. 33

Bacteria 40–53, *50*, 65, *67*, 82–4, 103, 116
Barnard's Star 128, 190
bats *149*
bees *60*
Bell, I. 201
Bernal, J. D. 79
Berzelius, J. J. 78, 115
Bessel, F. W. 13
Big-Bang theory 19–20, *20*
biomachines 164
bit (binary digit) 65
black-body radiation 20, 205
Blum, H. 87

Bondi, H. 19
Borman, F. 167
Bracewell, R. N. 194
Brahe, Tycho 11
bristle-cone pines 48, *49*
Bruno, Giordano 10

Cacti *42*
Calvin, M. 77, 87, 134
Cambrian period 27, 85
canals on Mars 101, *102*
Capek, K. 162
carbon 24, *24*, 138–41
carbon dioxide 89, 93
carbonaceous chondrites 115–16, 205
carnivorous plants 38, *39*, 151
catalyst 31, 205
cells 25, 82
α-Centauri 190–1
τ-Ceti *190*, 191, 197
chemoautotrophs 44, 205
chlorophyll 116
chloroplasts 82, *85*
chromatography 120–2, 205
chromosomes 28, 29, 33, 205
civilizations, extraterrestrial 190, 192
Claus, G. 115
Clements, H. 135
coacervates 78–80
Cocconi, G. 189, 204
comets 8, *10*, 78
complexity of life 62–70
Connington, J. J. 184
contamination 180–7
Copernican system 10, *12*
Cowie, D. 180
Crab Nebula *17*, 196
Crick, F. H. C. 34
critical temperature 135, 206
crystals 57, *57*, 64, *65*, 131
Curtis, H. D. 14
61-Cygni 128

Darwin, C. 27, 73
Deimos 114
deoxyribonucleic acid (DNA) 28, 33–7, *33–5*, 65–7, *69*, 155, 206
Devonian period 110
Dickson, G. R. 161
Digges, Thomas 11
distances to stars and galaxies 13–15
dolphins 152, *153*, 189
Doppler effect 198, 206
dormancy 48–53
Drake, F. D. 196, 198
Duchesne, J. 116
Dyson, F. 118, 176

Earth 9, 86, 90, *96*, *167*
  age of 20
  atmosphere 97
  threat to 182
ecology 39
Edie, L. C. 199
edelweiss *40*
Einstein, A. 23
electrons 23–4, *24*
elements 24
elephant 152
energy for life 44
energy for space travel 168–70
enzymes 31, *63*, 206
eobionts 78–81, 206
Eratosthenes 10
ε-Eridani 190–1, *190*, 197
escape velocity 90, 169, 206
eucaryotes 82, *83*, 206
evolution 58–60, *58–9*, 143–50, 161
exobionts 131, 146, 183, 206

Fictional monsters 114, *145*
Firsoff, V. A. 136, 143
Fisher, R. A. 29
flame 56, 57–8, 62
Flying Saucers 201
fossils 27, 53, *81*, 85
Fox, W. S. 79
Freudenthal, H. 200

fruit fly 28, *29*
Fuller, Buckminster 180
fungi, 38, *38*, 43
Fyfe, H. B. 202

Gagarin, Y. 167, *167*
galaxies 14–15, *15*, 18, 129, 180
Galileo 11
Ganymede 114
Gauss, C. F. 189
gene 28, 33–7, 206
genetic code 34–7, *36*, 80
Gold, T. 19, 74
Good, I. J. 156, 165
Good's Law 157, 161
gravitation 11
    and formation of stars 21–2
    and life 45–6, 147, 172
    and space travel 168, 172
greenhouse effect 93, 97, 106
growth 55
Gulliver 123, *123*
*Gymnarchus* 149, *149*

Haldane, J. B. S. 76, 87, 133, 136
Harvey, W. 26
heat resistance 42, 48–50, 142
helix 34–5, *34*
Henderson, L. J. 134
Hinton, H. E. 50
Holmes, D. C. 197
hormones 146
hot springs *41*, 42
Hoyle, F. 19, 59, 146, 202, 204
Hubble, E. P. 14
Huxley, Julian 150
hydrocarbons 94, 135
hydrogen 22, 78, 89
    bonds 134
    radio emission 17, 194
hydrogen cyanide 78, 106, 138

Immunity 184
information content 62–70, 122
intelligence 144, 150–65
interstellar communication 189–207
    codes for 198, 201
    languages for 200
ionization 134–6, 206
isotope 28, 206

Jansky, K. G. 17
Jung, C. G. 204
Jupiter 90, *104*, 105–6, 136, 137

Kluyver, A. J. 31

Lasers 193, *194*, 206
Laws of Robotics 163
Lederberg, J. 122, 180, 184
Leeuwenhoek, Anton van 25
Lemaître, G. 19
Lewis, C. S. 144
lichens 38, *38*, *43*, 44
life 25, 55, 70, 156
    alien 124–9
    alternative biochemistries of 131–43
    machines and 161, 164
    nature of 55–71
    on other planets 95, 100
    origin of 26, 73–81
    terrestrial 25–53
    unity of 31
    variety of 38
life-detection apparatus 118–24, *121–4*
life zones *117*
light-year *13*, 206
Lilly, J. C. 189
limonite 99
Linnaeus, C. 27
longevity 48
lotus 51
Lovell, J. 167
Lovelock, J. E. 121
Lowell, P. 101
lupin 51, *52*

Machines as life forms 156
Main-sequence 207
Malpighi, M. 26
mammals 86, 155
man 86, 144, 203
Mars 90, 97–103, *98–9*, *101–3*, 182
Mars jars 103
megacycles 207
Mendel, G. 28, *28*
Mercury 90–2, *91*
metabolic map *30*, 31, 65
metastable equilibrium 61, 122
meteorites 20, 78, 112
    organic matter in 114–18, 199
methane 76, 105–7
micro-organisms (see also algae, bacteria, viruses) 42, 74, 82, 115, 181, 207
Milky Way 11, 15, 125, *127*
Miller, S. L. *76*, 77

moon 108–13, *108–12*
Moore, P. 102
Morgan, T. H. 28
Morrison, P. 103, 189, 194, 204
Multivator *124*
Murray Meteorite 116
mutation 29, 33, 207

Nagy, D. 115
natural selection 27–9, 58–60
nebulae (see also galaxies) *17*, *21*, 207
Neolithic period 10
Neptune 90, 107, *107*
Newton, Isaac 11, *13*
Nirenberg, M. 36
nitrogen 76, 89, 93
nucleic acids (see also deoxyribonucleic acid) 31–7, 79, 131, 207
nucleotides 33, 36
nucleus 207
    atomic 23–4
    cell 28

Ødum, S. 51
Oparin, A. I. 74, *74*, 76, 78
optical activity 116, 121, 133
Order of the Dolphin 189
organization 61
Orgueil meteorite 115–17, 199
origin of life 73–81
Orion Nebula *21*
Oro, J. 78
oxygen 81, 121
Ozma, Project 196
ozone layer 48, 85, 97

Palaeolithic period 8
panspermia hypothesis 74, 112
parsec 14, 207
Pasteur, L. 27, *27*
Penrose, L. S. 69
Penrose models 69–70, *71*
Perrett, C. J. 70
pH 207
Phobos 114
photon 23
photosynthesis 82, 207
Pimental, C. G. 135
Pirie, N. W. 131
planets, 8, 89–108, 126, 180
Pluto 90, 107, *107*
polar caps 99–100, *99*

polymers 132, 208
*Polypedilum* 50
Ponnamperuma, C. 78
porpoise *148*, 149, *188*
potassium-40 20, 53
pressure, effect on life of, 45
Preyer, W. 74
procaryote 82, 208
proteins (see also enzymes) 31–2,
  *32, 37,* 51, 131, 208
proton 24
psychozoa 150, 208
Ptolemaic system 10, *12*
pulsars *19*, 196
Pythagoras 10

Quantum mechanics 23
quarantine 187
quark 24
quasar 18, *18*

Radiation 23, 194
  black-body 20, 205
  gamma and X-ray 20, 46, 48
  hydrogen 17, 194
  infra-red 150
  radio-wave 20, 148
  resistance of life to 46, 53
  ultraviolet 48
radio astronomy 16
radio sources 17–19, *18, 19,* 196
radio telescope *16, 195,* 196, *199*
radio transmission 193, 202
radioactivity 20, 53, 200, 208
rattlesnake 150, *150*
recycling in life support systems
  *174,* 175–6
red shift 15, *15,* 208
red tides 184, *185*
Redi, Francesco *26*

replication 35
reproduction 58, 161
Reynolds, J. E. 141
ribonucleic acid 36
ribosomes 37
Ritchey, G. W. 14
robots *51,* 59–60, 156, *160,* 162–4
rockets *88,* 167–77, *168–71*
roentgen 46, 208
Rutherford, Lord Ernest, 23

Sagan, C. 74, 95, 116, 192, 202
salinity 44
Sänger, E. 177
Sanger, F. 31
satellites 108, 113, 167
Saturn, 90, *105,* 106, *136*
Schiaparelli, G. 101
seeds 50
sequoia *54*
Shapley, H. 118
Shklovskii, I. S. 114
silica 141
silicates 94
silicon *141,* 141–3
silicones 142, *142*
Simpson, G. G. 87
Sinton, W. M. 100
soils 43, 59
solar flares 112, *113,* 173
solar system 8, 15, 22, 89, 167
solvents for life systems 134–8
sonar 149
space probes *94, 119,* 194
space travel 167–187
spacecraft 170–177, *172–3, 177*
spectroscopy 90, 120, 208
spontaneous generation 26
spores 48–52, *50,* 74–5
Stapledon, O. 146, 154, 165

stars 8–15, 118, 176
  nearest 190–1
Steady-State theory 19–20, *20*
stratosphere 208
sun 10–13, 75, 118
supernova 11, 14, 17, 22, 208
survival of life 41–8
symbionts 38, 84, 164, 208

Temperature extremes for life 41–2,
  48–9
Thales 10
Thompson, J. J. 23
time dilatation 177–80
Titan 114
tortoises 48
Triton 114

UFOs 201–3
ultraviolet light 48
universe, origin and age of 18–20
Uranus 90, 107, *107*
Urey, H. C. 77, 87, 118

Valency 24, 140, 208
Van Allen belts 97, 173
Venus 90, 93–6, *93, 95,* 182
Venus Fly Trap 151, *151*
viruses 66–9, *67, 68,* 185
Vishniac, W. 123

Water 43, 77, 98, 117, 133–5
Watson, J. D. 34
Wells, H. G. 59, 144, 182
whales 149, 152–3
Williams W. T. 162
Wolf Trap 123, *123*
Wyndham, J. 155

Xenobionts 131, 208